Principles of Cancer Treatment and Anticancer Drug Development

Wolfgang Link

Principles of Cancer Treatment and Anticancer Drug Development

 Springer

Wolfgang Link
Instituto de Investigaciones Biomédicas
"Alberto Sols" (CSIC-UAM)
Madrid, Spain

ISBN 978-3-030-18721-7   ISBN 978-3-030-18722-4 (eBook)
https://doi.org/10.1007/978-3-030-18722-4

© Springer Nature Switzerland AG 2019
This work is subject to copyright. All rights are reserved by the Publisher, whether the whole or part of the material is concerned, specifically the rights of translation, reprinting, reuse of illustrations, recitation, broadcasting, reproduction on microfilms or in any other physical way, and transmission or information storage and retrieval, electronic adaptation, computer software, or by similar or dissimilar methodology now known or hereafter developed.
The use of general descriptive names, registered names, trademarks, service marks, etc. in this publication does not imply, even in the absence of a specific statement, that such names are exempt from the relevant protective laws and regulations and therefore free for general use.
The publisher, the authors and the editors are safe to assume that the advice and information in this book are believed to be true and accurate at the date of publication. Neither the publisher nor the authors or the editors give a warranty, expressed or implied, with respect to the material contained herein or for any errors or omissions that may have been made. The publisher remains neutral with regard to jurisdictional claims in published maps and institutional affiliations.

This Springer imprint is published by the registered company Springer Nature Switzerland AG
The registered company address is: Gewerbestrasse 11, 6330 Cham, Switzerland

This book is dedicated to my wife, Ana

Preface

This book explains how current medicines against cancer work and how we find new ones. This is in a nutshell the content of the current book which is based on advanced classes in a master's degree program on "Molecular Mechanisms of Cancer." This postgraduate program includes learning modules on cell division and cell cycle, signal transduction, oncogenetic and oncogenomics, genetic toxicology, cell differentiation and stem cells, microenvironment, tumor progression, and clinical oncology. The learning module defeating cancer includes two main parts, one on cancer therapy and the second on drug development. The first part of Chap. 1 provides an introduction to the major problems we face when we treat malignant tumors in humans, setting the stage for Chap. 2 with a systematic overview over the current options of cancer therapy such as surgery, radiation, immunotherapy and chemotherapy and their corresponding modes of action. In particular, we focus on standard chemotherapeutic and targeted drugs and explain how they work. In Chap. 3, we cover intrinsic and acquired resistance mechanisms which represent the main obstacle to improve the clinical outcome of current cancer therapies. Chapter 4 is dedicated to the discovery and development of novel anti-cancer drugs. The development of new medicines to treat cancer nowadays is based on our growing understanding of the disease. In order to understand this paradigm shift, we discuss how medicines have been discovered and developed in the past. The structure of Chap. 2 reflects the development process of a new drug beginning with the identification and validation of a therapeutic target, the identification of an inhibitor of the target and its subsequent preclinical and clinical development. Finally, we describe the approval process by regulatory authorities. Chapter 5 provides a critical examination of the economic aspects of our current system of developing new medicines and its impact on our societies and on future drug research. At the end of each chapter, a section with thought questions and further reading has been added.

Madrid, Spain   Wolfgang Link

Acknowledgements I thank Javier Pérez (Scientific Photography and Drawing Facility, Instituto de Investigaciones Biomédicas "Alberto Sols," who prepared all the figures. I am grateful for the support, inspiration, and patience of my family—my wife Ana and our children, Lucia, and Pablo.

Contents

1 Introduction 1
   1.1 Introduction 1
      1.1.1 Why Haven't We Cured Cancer yet? 1
      1.1.2 Cancer Is Complex 2
      1.1.3 Where Do We Stand in Cancer Therapy? 4
   References 5

2 Cancer Therapy 7
   2.1 Mechanical Treatment 7
      2.1.1 Surgery 7
   2.2 Physical Treatments 9
      2.2.1 Radiation Therapy 9
      2.2.2 Photodynamic Therapy 13
      2.2.3 Hyperthermia 13
   2.3 Chemical Treatment 14
      2.3.1 Chemotherapy 14
      2.3.2 Targeted Therapy 32
   2.4 Biological Treatment: E.g. Immunotherapy and Oncolytic Viruses 59
      2.4.1 Immunotherapy 60
      2.4.2 Oncolytic Viruses 70
   References 72

3 Cancer Drug Resistance 77
   3.1 Mechanisms Upstream of the Molecular Target 78
   3.2 Mechanisms at the Level of the Molecular Target 80
   3.3 Mechanisms Downstream of the Molecular Target 81
   References 84

4	**Drug Discovery and Development**	87
	4.1 Target Identification	89
	4.2 Target Validation	92
	4.3 Lead Discovery and Optimization	98
	4.4 Pre-clinical Drug Development	118
	4.5 Approval Process for a New Drug	131
	References	134
5	**Economic and Social Implications of Modern Drug Discovery**	137
	References	138
Index		141

Abbreviations

ABC	ATP-binding cassette
ADME	Absorption, distribution, metabolism, excretion
AIDS	Acquired immune deficiency syndrome
ALCL	Anaplastic large-cell lymphoma
ALK	Anaplastic lymphoma kinase
ALL	Acute lymphoblastic leukemia
ATM	Ataxia-telangiectasia mutated
BCC	Basal cell carcinoma
BER	Base excision repair
BLA	Biologic license application
BRCA1/2	BReast CAncer genes 1 and 2
CAFs	Cancer-associated fibroblasts
CAR	Chimeric antigen receptors
CDX	Cell line-derived xenograft
CHO	Chinese hamster ovary
CLL	Chronic lymphocytic leukemia
CML	Chronic myelogenous leukemia
CRC	Colorectal cancers
CRISPR	Clustered regularly interspaced short palindromic repeats
CTCL	Cutaneous T-cell lymphoma
CTLA-4	T-lymphocyte-associated protein 4
CYP450	Cytochromes P450
DHFR	Dihydrofolate reductase
DLBCL	Diffuse large B-cell lymphoma
DLTs	Dose-limiting toxicities
DNA-PK	DNA-dependent protein kinase
DSB	Double-strand breaks
EC	Effective concentration
ECM	Extracellular matrix
EGFR	Epidermal growth factor receptor

EMA	European medicines agency
EMT	Epithelial-to-mesenchymal transition
ERα	Estrogen receptor-α
Fab	Fragment antigen binding
Fc	Fragment crystallization
FDA	Food and Drug Administration
FGFR	Fibroblast growth factor receptor
FISH	Fluorescence in situ hybridization
FL	Follicular lymphoma
GARTF	Glycinamide ribonucleotide formyltransferase
GEJ	Adenocarcinoma of the gastroesophageal junction
GEM	Genetically engineered mice
GIST	Gastrointestinal stromal tumor
GLI	Glioma-associated transcription factor
GLP	Good laboratory practice
GM-CSF	Granulocyte-macrophage colony-stimulating factor
HATs	Histone acetyltransferases
HCS	High-content screening
HDACs	Histone deacetylases
HEK	Human embryonic kidney
Her2	Human epidermal growth factor receptor 2
hERG	Ether-à-go-go-related gene
HNC	Head and neck cancer
HNSCC	Head and neck squamous cell carcinoma
HPLC	High-performance liquid chromatography
HRR	Homologous recombination repair
HSV1	Herpes simplex virus 1
HTS	High-throughput screening
IAPs	Inhibitor of apoptosis proteins
IC	Inhibitory concentration
IHC	Immunohistochemistry
IL2rgnull	IL2 receptor common gamma chain
IND	Investigational New Drug
IP	Intellectual property
IRB	Institutional review board
LC/MS	Liquid chromatography/mass spectrometry
LINAC	Linear accelerator
LOF	Loss of function
mAb	Monoclonal antibody
MAPK	Mitogen-activated protein kinase
MCC	Merkel cell carcinoma
MCL	Mantle cell lymphoma
MDR1	Multi-drug resistance protein 1
MGMT	Methylguanine methyltransferase
MHC	Major histocompatibility complex

miRNA	MicroRNA
MM	Multiple myeloma
MTD	Maximum tolerated dose
mTOR	Mammalian target of rapamycin
NCE	New chemical entity
NCI	National Cancer Institute
NDA	New Drug Application
NER	Nucleotide excision repair
NHL	Non-Hodgkin's lymphoma
NME	New molecular entity
NSCLC	Non-small-cell lung cancer
ORR	Overall response rate
OS	Overall survival
PAMP	Pathogen-associated molecular pattern
PAMPA	Parallel artificial membrane permeability assay
PAP	Prostatic acid phosphatase
PARP	Poly(ADP-ribose) polymerase
PD-1	Programmed death-1
PDGFR	Platelet-derived growth factor receptor
PDT	*Photodynamic therapy*
PDX	Patient-derived xenograft
PFS	Progression free survival
PI3K	Phosphatidylinositide 3-kinases
PIP3	Phosphatidylinositol-3,4,5-trisphosphate
PRR	Pattern recognition receptor
PTCH1	Patched1
PTCL	Peripheral T-cell lymphoma
PTEN	Phosphatase and tensin homolog
RCC	Renal cell carcinoma
RNAi	RNA interference
ROS	Reactive oxygen species
RR	Ribonucleotide reductase
RTK	Receptor tyrosine kinase
SAR	Structure–activity relationship
SCFR	Mast/stem cell growth factor receptor
SCID	Severe combined immunodeficiency
Shh	Sonic hedgehog
shRNA	Small hairpin RNA
siRNA	Small interference RNA
SMO	Smoothened
SPR	Structure–property relationships
STD	Severe toxicity dose
SUFU	Suppressor of fused
TAM	Tumor-associated macrophage
TCL	T-cell lymphoma

TdP	Torsades de pointes
TF	Transcription factor
TS	Thymidylate synthase
TTP	Time to progression
VEGF	Vascular endothelial growth factor
VEGFR	Vascular endothelial growth factor receptor

Chapter 1
Introduction

1.1 Introduction

We are currently witnessing a paradigm shift in the way we find new medicines to treat cancer based on our growing understanding of the disease. This book explains the mechanisms underlying the action of current cancer therapies and describes how we can translate our accumulated knowledge into innovative therapies.

1.1.1 Why Haven't We Cured Cancer yet?

Smallpox, a devastating viral disease was declared eradicated in 1980 following a global immunization campaign (Behbehani 1983). Modern medicine turned AIDS from a deadly disease into a chronic, manageable condition in the late nineties of the last century (Deeks et al. 2013). We even improved the treatment of cardio-vascular diseases to a point that cancer whose incidence continues to rise, both in high and low-income countries, will replaces these diseases as the major cause of death within the near future in many regions of our planet (Allemani et al. 2018). Is it really so much harder to find a solution for cancer? The short answer is yes, it is. The more extended discussion on this issue below tries to provide different conceptual frameworks to understand the problems associated with treating cancer (Table 1.1). First of all, cancer is a disease of our own cells. This implies that our immune system which eliminates foreign pathogens very efficiently, has not been optimized to fight cancer and that it is difficult to find therapeutic drug targets that are only present in cancer cells and absent in our normal cells. Second, cancer is not a single disease, but many different ones and for some of them there are now very successful treatment options available. Here it is important to note, that there is not a complete lack of success stories in the cancer field. And that cancer is very complex. Nowadays, we distinguish more than a hundred different cancer types. Third and may be the most critical obstacle for more effective anti-cancer therapies is clonal

Table 1.1 List of obstacles to successful cancer therapy

No	Obstacle
1	Disease of our own cells
2	Hundred different cancer types
3	Clonal heterogeneity
4	Complex interaction with the microenvironment
5	Cancer stem cells
6	Experimental models fail to represent human tumors

heterogeneity as a substrate of a Darwinian process of environmental selection (McGranahan and Swanton 2015; Turajlic et al. 2019). Cancer treatment represents a selective pressure which will invariably lead to the appearance of resistant populations of cancer cells (McGranahan and Swanton 2015). Fourth, yet another challenge comes from the complex and dynamic interactions of cancer cells with non-cancer cells in their environment (Bhowmick et al. 2004; Mueller and Fusenig 2004; Polyak et al. 2009). These interactions might help to feed the tumor with oxygen and nutrients and promote evasion of cancer cells from programmed cell death or destruction by the immune system. Fifth, cancer stem cells are thought to be resistant to a broad variety of cancer therapies and might cause relapse (Batlle and Clevers 2017; Reya et al. 2001). The cancer stem cell theory may explain why anti-cancer therapies often fail to produce an improved clinical outcome even if the bulk of the tumor has shrunk beyond detection. And sixth, last but not least, most of our current models based on cultured cells or animals that we use to study cancer fail to represent the complexity displayed by a tumor in a human body (Hoelder et al. 2012; Hutchinson and Kirk 2011; Ruggeri et al. 2014). Hence many experimental therapies seem to work in our model systems, but turn out to have limited efficiency against human tumors in the real world.

1.1.2 Cancer Is Complex

People would never think of malaria, tuberculosis and cholera as being the same disease despite the fact that they all three are infectious diseases. However, a common misconception is, that the word 'cancer' defines a single disease. As the terms infectious diseases or cardiovascular diseases, cancer describes a group of different diseases which are all characterized by an uncontrolled proliferation of cells which have the potential to invade or metastasize (Benz 2017; Berman 2004). There are more than a hundred types of cancer that might be classified into five groups: Carcinomas, Sarcomas, Neuroectodermal malignancies, Hematopoietic malignancies and atypical tumors (Table 1.2).

The majority of human tumors arise from epithelial tissues that line the walls of cavities and channels. A cancer that develops from epithelial cells is called carcinoma. The main functions of epithelia are protection and secretion. Tumors that

1.1 Introduction

Table 1.2 List of cancer types and corresponding incidence and mortality

Cancer type	Incidence (%)	Mortality (%)
Carcinoma	>85	>85
Sarcoma	~1	<1
Hematopoietic malignancies	~6.5	~7
Neuroectodermal malignancies	~1.3	~2.5
Atypical tumors	~7	n.d.

arise from protective layers of epithelial cells are called squamous cell carcinomas, whereas adenocarcinomas are formed by cells that secrete substances into the ducts (Dotto and Rustgi 2016). Sarcomas are nonepithelial tumors which arise from connective tissues such as adipose tissue, cartilage or bone (Doyle 2014). Hematopoietic malignancies arise from the blood-forming tissues and are classified into three major groups, leukemias, cancers of white blood cells, lymphomas tumors formed by B and T cells that form solid tumor masses and myelomas formed by antibody producing plasma cells (Taylor et al. 2017). As cells of the central and peripheral nervous systems derive from the ectoderm, tumors formed by these cells are called neuroectodermal tumors (Kleihues et al. 2002). Melanomas, small-cell lung carcinomas, teratomas and some other tumors do not fit into the five major tumor types described above and are classified as atypical tumors. In rare cases (<4%) it is not possible to identify the original tissues of a tumor which is then classified as a cancer of unknown primary (CUP) (Pavlidis and Fizazi 2009). The classification of tumor types as described above is based on identifying the tissue of origin according to histopathological criteria. As we increase our knowledge on the genetic and epigenetic alterations present in these tumors, we realize that the histopathological classification of tumors is not always useful to guide therapeutic interventions (Golub et al. 1999). Understanding the molecular mechanism underlying the disease becomes increasingly important to choose therapeutic options and predict clinical response. Scientists have investigated the molecular mechanisms of cancer for the last forty years and have accumulated a significant amount of knowledge about cancers and their causes. Molecular cancer research dates back to the early seventies of the last century and led to the discovery of oncogenes and tumor suppressor genes as well as the identification of complex signalling circuits whose alterations drive tumor formation and progression. However, the pace of translation of knowledge into the clinic was slow and sometimes disappointing. We are now beginning to witness significant improved treatment options for several cancer entities based on the molecular understanding of cancer. In Chap. 4 we will learn about strategies aimed at systematically converting our knowledge into validated therapeutic targets as a starting point of drug discovery projects.

1.1.3 Where Do We Stand in Cancer Therapy?

Again, as in the case of infectious diseases or other complex groups of disorders, there won't be a singular cure for cancer. In fact, some of the cancers mentioned above are readily curable nowadays, whereas others still have a very poor clinical outcome. The approval of the chemotherapeutic agent cisplatin in 1978 for the treatment of testicular cancer increased the survival rate to over 95% (Masters and Koberle 2003). The survival rate of pediatric patients with acute lymphoblastic leukemia (ALL), the most common form of pediatric leukemia for which several treatment options have become available including chemotherapy regimens and hematopoietic stem cell transplants is way above 80%. Women whose HER2-positive or estrogen-receptor-positive breast cancer responds to Trastuzumab or tamoxifen/aromatase inhibitors, respectively have a very high probability of being cured. The targeted drug imatinib, FDA-approved in 2001, leads to remissions in the majority of patients with chronic myelogenous leukemia (CLL) and converted this rapidly fatal disease into a chronic condition. The spectacular success of immune checkpoint inhibitors in a subset of patients with advanced melanoma has generated huge expectations in cancer immunotherapy, a strategy that harnesses the body's immune system to fight cancer (Sharma and Allison 2015). On the other hand, we have failed to significantly improve the treatment options for the major epithelial tumors such as lung, colon and stomach cancers. It is true that mortality is generally improving, even for some of the more lethal cancers such as lung and liver cancer. However, increasing survival rates for these tumors seem to be largely due to better prevention such as smoking cessation and earlier detection through screening. Accordingly, survival rates often depend significantly where you live (Allemani et al. 2018). For some cancers such as pancreatic cancer, 5-year survival remains extremely low worldwide ranging between 5 and 15%. It is quite safe to predict that cancer will be part of the human condition for the foreseeable future, but we can improve cancer prevention, accelerate diagnosis and develop safer and more efficient therapies when it appears. The latter will be extensively discussed in the second part of this textbook. But let's first have a look on the available options to treat cancers.

Thought Questions

1. The majority of human tumors arise from:
   - (A) Skin
   - (B) Epithelial tissues
   - (C) Blood forming cells
   - (D) Cells of the central and peripheral nervous systems
   - (E) Connective tissues

2. Which of the following statements about sarcomas is true:

   (A) Sarcomas are epithelial tumors derived from the basal lamina
   (B) Sarcomas are nonepithelial tumors derived from tissue of ectodermal origin
   (C) Sarcomas are nonepithelial tumors derived from various connective tissues
   (D) Sarcomas are nonepithelial tumors derived from blood forming cells
   (E) Sarcomas are rare tumors derived from neuromuscular tissue

3. What is a CUP (cancer of unknown primary)?

References

Allemani C et al (2018) Global surveillance of trends in cancer survival 2000–14 (CONCORD-3): analysis of individual records for 37 513 025 patients diagnosed with one of 18 cancers from 322 population-based registries in 71 countries Lancet 391:1023–1075. https://doi.org/10.1016/s0140-6736(17)33326-3

Batlle E, Clevers H (2017) Cancer stem cells revisited. Nat Med 23:1124–1134. https://doi.org/10.1038/nm.4409

Behbehani AM (1983) The smallpox story: life and death of an old disease. Microbiol Rev 47:455–509

Benz EJ Jr (2017) The Jeremiah Metzger lecture cancer in the twenty-first century: an inside view from an outsider. Trans Am Clin Climatol Assoc 128:275–297

Berman JJ (2004) Tumor classification: molecular analysis meets Aristotle. BMC Cancer 4:10. https://doi.org/10.1186/1471-2407-4-10

Bhowmick NA, Neilson EG, Moses HL (2004) Stromal fibroblasts in cancer initiation and progression. Nature 432:332–337. https://doi.org/10.1038/nature03096

Deeks SG, Lewin SR, Havlir DV (2013) The end of AIDS: HIV infection as a chronic disease. Lancet 382:1525–1533. https://doi.org/10.1016/S0140-6736(13)61809-7

Dotto GP, Rustgi AK (2016) Squamous cell cancers: a unified perspective on biology and genetics. Cancer Cell 29:622–637. https://doi.org/10.1016/j.ccell.2016.04.004

Doyle LA (2014) Sarcoma classification: an update based on the 2013 World Health Organization classification of tumors of soft tissue and bone. Cancer 120:1763–1774. https://doi.org/10.1002/cncr.28657

Golub TR et al (1999) Molecular classification of cancer: class discovery and class prediction by gene expression monitoring. Science 286:531–537

Hoelder S, Clarke PA, Workman P (2012) Discovery of small molecule cancer drugs: successes, challenges and opportunities. Mol Oncol 6:155–176. https://doi.org/10.1016/j.molonc.2012.02.004

Hutchinson L, Kirk R (2011) High drug attrition rates–where are we going wrong? Nat Rev Clin Oncol 8:189–190. https://doi.org/10.1038/nrclinonc.2011.34

Kleihues P, Louis DN, Scheithauer BW, Rorke LB, Reifenberger G, Burger PC, Cavenee WK (2002) The WHO classification of tumors of the nervous system. J Neuropathol Exp Neurol 61:215–225. (Discussion 226–219)

Masters JR, Koberle B (2003) Curing metastatic cancer: lessons from testicular germ-cell tumours. Nat Rev Cancer 3:517–525. https://doi.org/10.1038/nrc1120

McGranahan N, Swanton C (2015) Biological and therapeutic impact of intratumor heterogeneity in cancer evolution. Cancer Cell 27:15–26. https://doi.org/10.1016/j.ccell.2014.12.001

Mueller MM, Fusenig NE (2004) Friends or foes—bipolar effects of the tumour stroma in cancer. Nat Rev Cancer 4:839–849. https://doi.org/10.1038/nrc1477

Pavlidis N, Fizazi K (2009) Carcinoma of unknown primary (CUP). Crit Rev Oncol Hematol 69:271–278. https://doi.org/10.1016/j.critrevonc.2008.09.005

Polyak K, Haviv I, Campbell IG (2009) Co-evolution of tumor cells and their microenvironment. Trends Genet 25:30–38. https://doi.org/10.1016/j.tig.2008.10.012

Reya T, Morrison SJ, Clarke MF, Weissman IL (2001) Stem cells, cancer, and cancer stem cells. Nature 414:105–111. https://doi.org/10.1038/35102167

Ruggeri BA, Camp F, Miknyoczki S (2014) Animal models of disease: pre-clinical animal models of cancer and their applications and utility in drug discovery. Biochem Pharmacol 87:150–161. https://doi.org/10.1016/j.bcp.2013.06.020

Sharma P, Allison JP (2015) The future of immune checkpoint therapy. Science 348:56–61. https://doi.org/10.1126/science.aaa8172

Taylor J, Xiao W, Abdel-Wahab O (2017) Diagnosis and classification of hematologic malignancies on the basis of genetics. Blood 130:410–423. https://doi.org/10.1182/blood-2017-02-734541

Turajlic S, Sottoriva A, Graham T, Swanton C (2019) Resolving genetic heterogeneity in cancer. Nat Rev Genet. https://doi.org/10.1038/s41576-019-0114-6

Further Reading

DeVita VT Jr (2015) DeVita, Hellman and Rosenberg's cancer: principles and practice of oncology, 10th edn. Wolters Kluwer Health, Philadelphia, USA, pp 1241–1313

Hanahan D, Weinberg RA (2000) The hallmarks of cancer. Cell 100:57–70

Weinberg R (2014) The biology of cancer, 2nd edn. Garland Science

Chapter 2
Cancer Therapy

Cancer is probably as old as human life on our planet and the price we pay for the ability to replace and repair tissues throughout our life span. Many cancers are avoidable, though, through changes in lifestyle for example by quitting smoking, eating low fat diet or cancer screening and early detection. However, not all cancers can be prevented. Many millions of patients will continue to be diagnosed with cancer every year worldwide. In order to reduce cancer mortality, reduction of incidence and improvement of survival are both necessary. There are many different treatments for cancer available depending on the type and stage of the tumor. The most common treatment modalities for cancer are surgery, radiation, immunotherapy and chemotherapy. Many of these options can be combined sequentially or in parallel. The modalities to treat cancer can be classified according to different criteria such as type of target, mode of action, type of agent and others. Maybe the simplest way of dividing the treatments into subgroups is to classify them into mechanical treatments, physical treatments, chemical treatments and biological treatments (Fig. 2.1).

2.1 Mechanical Treatment

2.1.1 Surgery

Mechanical removal of the tumor by surgery is one of the oldest and still the most frequently used treatment option and is often the modality with the greatest potential benefit. Surgery is a medical procedure to remove, examine or repair tissue. It is a local treatment only useful for solid tumors that are contained in one area. Surgery is not used for liquid tumors such as leukemia or for cancers that have metastasized. Furthermore, surgery often fail to remove all malignant cells from the patient and therefore the tumor might grow back after some time. Hence, surgery is often combined with neoadjuvant or adjuvant chemotherapy (see Box 1 for the definition

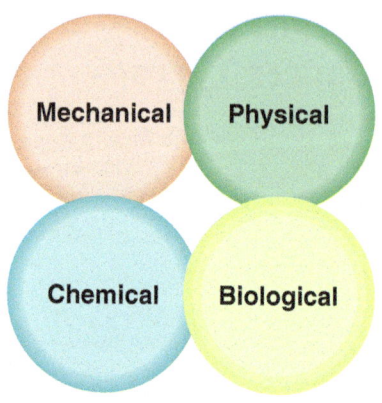

Fig. 2.1 Classification of modalities to treat cancer. Mechanical Treatment, e.g. surgery, physical treatment, e.g. radiotherapy, hyperthermia, chemical treatment, e.g. chemotherapy and targeted treatments, biological treatment, e.g. immunotherapy and oncolytic viruses

of curative, adjuvant, neoadjuvant therapeutic, or palliative treatment), radiotherapy, hormonal therapy and others. Surgery is not always aimed at curing cancer but can play a role in cancer prevention e.g. prophylactic mastectomy to reduce the risk of breast cancer, cancer diagnosis and staging e.g. removal of lymph nodes for examination, palliative care e.g. to relieve pain in patients with advanced cancer or post-treatment reconstruction of tissues. Surgery is an option only for those patients fit enough to tolerate surgical procedures and anesthesia. The advent of surgical anesthesia first introduced in 1846 and the development of antiseptic and aseptic methods represent major breakthroughs that paved the way for modern surgery. Technological progress, in particular the development of sophisticated imaging technology enabled minimally invasive surgery to remove tumors including colon, esophagus and bladder tumors through tubes.

> **Box 1**
> **Concepts: curative, palliative, adjuvant, neoadjuvant treatment**
> The terms curative, palliative, adjuvant, neoadjuvant can be applied to several different treatment modalities such as surgery, radiation therapy and chemotherapy. Chemotherapy is curative if its intention is to cure the cancer as a single modality. When chemotherapy is administered following the main treatment e.g. surgery, radiation or hormone therapy to handle the remaining malignant cells it is called adjuvant chemotherapy. Conversely, neoadjuvant chemotherapy is administered before the main treatment. Neoadjuvant chemotherapy is given to treat unrespectable lung, colorectal and breast cancer or to shrink a tumor before surgery. Palliative chemotherapy is not intended to cure the disease but rather to alleviate symptoms and to improve quality of life of a terminal patient.

2.2 Physical Treatments

Physical treatments against cancer try to destroy the tumor using radiation, light or heat and include radiation therapy (also called radiotherapy), photodynamic therapy and hyperthermia.

2.2.1 Radiation Therapy

Radiation can cause cancer and cure it. Radiation therapy has been employed to treat cancer for more than 100 years, soon after the discovery of X-rays in 1895 by the German physicist Wilhelm Röntgen. Nowadays, radiation therapy is one of the most common treatment modalities with about 50% of patients with localized tumors receiving radiation therapy in the course of their disease in developed countries (Delaney et al. 2005).

What Is Radiation?
Radiation is the emission or transmission of energy in the form of waves or fast traveling particles through space or matter in the form of heat, sound and light (Bolus 2017). Radiation is produced by the interaction of a particle with matter or the radioactive decay of an unstable atom (Georg and Thwaites 2017; Lee et al. 2015). We distinguish electromagnetic radiation that consists of photons that have energy, but no mass or charge and particle radiation based on particles that have mass and energy, with or without electric charge. Electromagnetic radiation includes radio waves, microwaves, visible light, ultraviolet light, X-rays, and gamma rays. Examples of particle radiation are alpha particles, protons, beta particles, and neutrons (Fig. 2.2).

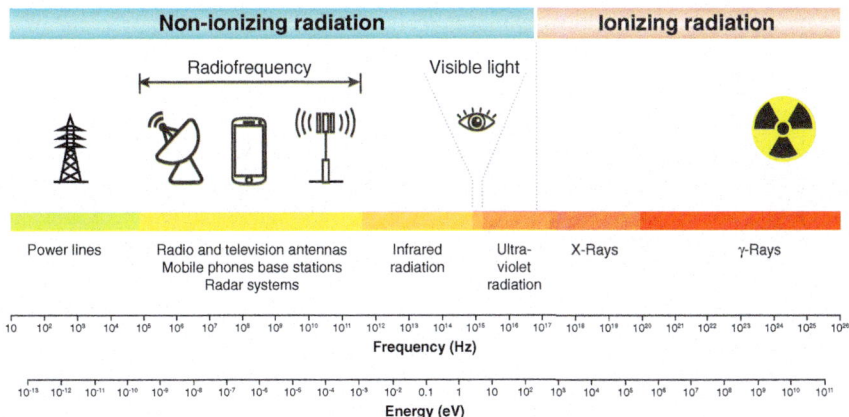

Fig. 2.2 Electromagnetic radiation. The spectrum of electromagnetic radiation represents the possible frequencies. The sources of non-ionizing and ionizing radiation are represented according their according wavelength

Depending on the energy and the ability to penetrate matter, radiation can break atoms to generate ions. Ions are electrically charged atoms or molecules formed when atoms loose or gain electrons. Radiation with sufficient energy to remove an electron from an atomic orbital, is called ionizing radiation. Ionizing radiation includes X-ray, gamma rays, alpha particles, and beta particles. Radiation therapy mainly employs ionizing radiation most frequently X-rays. On the other hand, radio waves, microwaves, and visible light waves have enough energy to causes atoms to move inside a molecule, but they are not capable of generating ions. Therefore, this kind of radiation is called non-ionizing radiation which is not harmful.

Radiation Can Be Administered from Outside or Inside the Body or Systemically

Ionizing radiation in radiation therapy can be delivered from an external source, from a source introduced into the body or by systemic administration of radioactive drugs (Formenti and Demaria 2009). External beam radiation therapy is the most common type of radiation therapy to treat cancer including head and neck, breast, lung, colon, and prostate cancers. High energy X-rays are generated outside the body by a specialized equipment e.g. a linear accelerator (LINAC). Linear accelerators accelerate electrons which then collide with heavy metals to generate a tightly targeted high energy beam directed to the patient's tumor. Surface tumors such as skin cancer are treated with low-energy radiation which does not penetrate very deeply whereas high-energy radiation is used to treat deeper cancers. There are several types of external radiation therapy including three-dimensional conformal radiation therapy, Image guided radiation therapy, intensity modulated radiation therapy and proton beam radiation therapy all having specific advantages. Recently, proton beam radiation therapy is receiving a lot of attention as it causes less damage to healthy tissue. This type of external beam radiation therapy is based on protons, positively charged subatomic particles present in the nucleus of every atom. Due to their relatively large mass, they have little lateral side scatter in the tissue. Because of high energy protons are produced by a particle accelerator, proton beam radiation therapy is far more expensive than conventional radiation therapy.

In internal radiation therapy, also called **brachytherapy** (from Greek, brachy, short) the source of radiation is placed in or near the tumor allowing for the delivery of a high dose of radiation to the malignant tissue while reducing the exposure of surrounding tissue (Koukourakis et al. 2009). Brachytherapy is used as a treatment for cervix, uterus, vagina, rectum, prostate, breast, eye, and head and neck cancers. Brachytherapy is a local treatment in which a short-range radiation-source is enclosed into containers such as needles, wires, tubes or rods which prevent systemic exposure to radiation. Radiation source may stay in place for minutes, hours, days or permanently.

Conversely, in **systemic radioisotope therapy**, unsealed radioisotopes such as radioactive iodine or radioactively labelled monoclonal antibodies are delivered orally or by injection to treat tumors, systemically. Despite the systemic administration of radioactive drugs, these agents are specifically targeted to the tumor. Radioactive iodine-131 can be used to treat thyroid cancer because the thyroid gland is capable of concentrating iodide 30–50 times that of the circulating concentration. The radionuclides phosphorus-32, strontium-89 and samarium-153 localize to regions of enhanced bone turnover in damaged bones and have been approved to treat bone metastasis. Radioimmunotherapy uses monoclonal antibody conjugated with radioisotopes to guide the radiation source to specific cells. Ibritumomab tiuxetan (Zevalin) is an anti-CD20 monoclonal antibody conjugated to yttrium-90 or indium-111 which binds to the CD20 antigen on the surface of B cells but not B cell precursors. Ibritumomab tiuxetan has been approved to treat refractory non-Hodgkins lymphoma (Witzig et al. 2002).

Radiation Therapy Uses Ionizing Radiation to Kill Cancer Cells
Ionizing radiation breaks covalent bonds of molecules and generates ions in the cells of the tissues it passes through (Prise and O'Sullivan 2009). Therefore, ionizing radiation is used in radiation therapy to damage cancer cells. Radiation therapy damages many different cellular components such as proteins, lipids and nucleic acids. Experiments with microbeam radiating cytoplasm and nucleus independently showed the main target of radiation damage is the nucleus. Damage by radiation to cellular DNA is either caused by direct ionization of atoms which are part of the DNA molecule or by the production of free radicals which then induce strand crosslinks, base damages, single-strand and double-strand breaks (DSB) in the DNA. DSBs represent the most severe type of DNA damage (Fig. 2.3). The majority of cellular damage is caused indirectly through free radicals. Free radicals are atoms, molecules, or ions with unpaired electrons. Free radicals that involve oxygen are called reactive oxygen species (ROS). The free electron makes these agents chemically reactive

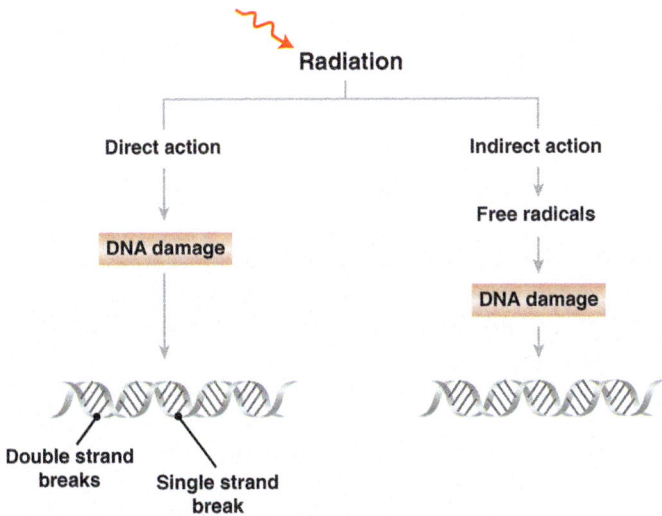

Fig. 2.3 Mode of action of radiation therapy. DNA Damage by radiation can be caused by direct ionization of atoms which are part of the DNA molecule or can be indirect by the production of free radicals which then induce DNA damage

capable of attacking the covalent bonds of other molecules including DNA. Free radical species may trigger chain reactions which can be highly damaging to cells leading to growth arrest at a cell cycle checkpoint to allow for DNA repair or apoptosis if the damage cannot be repaired.

Dividing Cells Are More Sensitive to Ionizing Radiation
Different cells display differing degrees of sensitivity to radiation therapy. Susceptibility of cells to ionizing radiation depends on several factors including cell cycle phase, metabolic activity, differentiation state and genetic factors (Pawlik and Keyomarsi 2004). Cells are most sensitive to ionizing radiation in the M phase of the cell cycle. Fast dividing cells are not capable of repairing the DNA damage induced by ionizing radiation fast enough and undergo apoptosis ensuring that the damaged DNA is not passed to the daughter cells. Hence, the use of radiation therapy in cancer is based on the rationale that rapidly proliferating cancer cells are more sensitive to DNA damage produced by this treatment modality than normal cells. However, cancer recurrence after successful radiation therapy is quite common suggesting the appearance of intrinsic or acquired radioresistance within a subpopulation of cancer cells. Mutations in genes involved in survival signalling, apoptosis, tumor suppression and DNA repair including PTEN, PI3 K, p53, ATM, BRCA1/2 and DNA-PK have been shown to influence cell vulnerability to radiation therapy. Importantly, cancer stem cells are thought to be less sensitive to ionizing radiation due to their quiescent state, enhanced DNA repair, upregulated cell cycle control, high level of free-radical scavengers and a microenvironment that supports cell survival. Therefore, even in the case where radiation therapy killed the bulk of the tumor cells, rare cancer stem cells remain and eventually repopulate the tumor (Vlashi and Pajonk 2015).

The Four "Rs" of Radiation Biology Affect the Response to Fractionated Radiation Therapy
Although the full radiation dose in external radiation therapy might be delivered in one session, most patients receive smaller doses called fractions over many weeks. The most important factors known to influence the cellular response to radiation therapy are repair, redistribution, repopulation and reoxygenation and have been coined the four "Rs" of radiation biology (Trott 1982). These factors are important to schedule radiation therapy. As fast dividing cancer cells are less efficient in repairing radiation induced DNA damage, radiation treatment can be scheduled to provide normal cells time to recover. Redistribution refers to the fact that tumors are unsynchronyzed populations of cells in different phases of the cell cycle and hence exhibiting differential sensitivity to radiation. Fractionation allows cancer cells which survived radiation treatment to cycle into a more sensitive phase of the cycle before the next fraction is delivered. Similarly, as oxygen enhances the damage caused by free radicals, hypoxic tumor cells are more resistant to radiation. By eliminating the nomoxic layer of cells by radiation treatment, the anoxic cells might get closer to the blood supply and reoxygenate between fractions rendering them cells more sensitive to the next dose. It takes some time for the tissues to repopulate the dead cells after

radiation treatment. In order to reduce side effects, fractionation might be useful to take advantage of differences in repopulation speed between healthy tissues and cancer cells.

Ionizing Radiation Damages Healthy Tissue and Might Induce Secondary Tumors

The effect of ionizing radiation is not limited to cancer cells, but damages also surrounding healthy, non-cancerous cells. Side effects from radiation depend on the dose and treated area of the body (Bentzen 2006). As sensitivity of healthy tissue is limiting the efficacy of radiation therapy efforts have focussed on increasing the tumor dose while minimizing the damage to surrounding tissue. This is also important to reduce the risk of secondary tumors (Dracham et al. 2018). Exposure of healthy tissue to ionizing radiation has been shown to result in a significant and dose dependent increase of secondary tumors near the first.

2.2.2 Photodynamic Therapy

Photodynamic therapy (PDT) utilizes a photosensitizing agents, light and oxygen to trigger a photochemical reaction that results in the generation of a highly-reactive singlet oxygen (Robertson et al. 2009). This high-energy form of oxygen can cause direct toxicity leading to cell death as well as damage the tumor vasculature and induce an inflammatory reaction. Porfimer sodium (Photofrin) is a photosensitizing molecule which has been approved for the palliative treatment of non-small cell lung and esophageal cancer (Overholt et al. 2005). Porfimer sodium is injected intravenously and distributes selectively to tumor tissues. After absorbtion of porfimer sodium, laser light is applied locally. Laser light can be directed through fiber optic cables to expose areas inside the body.

2.2.3 Hyperthermia

Hyperthermia is a therapy using a modest elevation of temperature in the range between 39 and 48 °C to specifically kill cancer cells (Gerweck 1985). Hyperthermia is one of the oldest known treatments for cancer mentioned in Egyptian papyrus 3000 BC and by Hippocrates but has been approved only in combination with other treatment modalities and is not widely available. The rational for using hyperthermia to treat cancer is that cancer cells might be more vulnerable to heat than normal cells due to low the oxygen level and acidic conditions found within tumor tissue. Hyperthermia has been shown to sensitize cancer cells to radiation therapy and the treatment with chemotherapy. There are many different approaches to use hyperthermia. Heat can be applied superficially or deep to the whole-body or locally, it can be used as sensitizer or ablator and the source of the heat can be external or internal. Hyperthermia treatment has encountered several obstacles including a lack of specificity towards cancer cells and non-homogenous heat dispersion throughout the tumor. Progress in nanotechnology and drug delivery technologies might address these issues. A promising approach is the use of functionalized magnetic nanoparticles specifically targeted to a tumor to remotely induce local heat by applying a magnetic field (Chatterjee et al. 2011).

2.3 Chemical Treatment

Chemical treatments of cancer are aimed at destroying the tumor using chemical agents such as cytotoxic chemotherapeutic drugs and targeted drugs. Chemical treatments constitute a systemic therapy introduced into the blood stream and are therefore in principle able to address cancer at any anatomic location in the body. It is often the only option for the treatment of metastatic disease where therapy has to be systemic throughout the patient. In this section, targeted drugs which are not considered chemotherapy will also be discussed.

2.3.1 Chemotherapy

The term chemotherapy has been introduced for the use of chemicals to treat diseases by the German-Jewish physician Paul Ehrlich who received the Nobel Prize for Medicine in 1908 (Strebhardt and Ullrich 2008). The origin of anti-cancer chemotherapy stretches back to the early 1940s when the toxic action of nitrogen mustard-based war gas on cells of the hematopoietic system was discovered (Box 2) (DeVita and Chu 2008).

> **Box 2**
> **Chemotherapy—a historical perspective**
> The origin of chemotherapy dates back to the early 1940s when the toxic action of mustard-based war gas on cells of the haematopoietic system was discovered. An important catalyzing event happened on 2 December 1943 during World War II. German bombers attacked Allied forces in the port of Bari in Italy, and mustard gas from a wrecked cargo ship was released. Military and civilian personnel were accidentally exposed to mustard gas. Many developed mustard poisoning and 83 died. Autopsy of the victims revealed profound lymphoid and myeloid suppression. In particular, the observation of a low white blood cell count strengthened the idea that mustard agents could be useful to treat cancers of the white blood cells. However, mustard gas is volatile and therefore difficult to handle in experiments. In order to generate a more stable compound, the American pharmacologists, Alfred Gilman and Louis Good-

2.3 Chemical Treatment

man substituted nitrogen for sulfur and tested the obtained nitrogen mustard (Mustine) in treating lymphoma. Nitrogen mustards are DNA alkylating agents that attach analkyl group (R-CH2) to the guanine base of DNA and interfere with DNA replication. Mustine was the first chemotherapeutic agent used to treat cancer and has its origin in chemical warfare. After World War II in 1948, the American pediatric pathologist Sidney Farber introduced the use of antifolates in children with acute lymphoblastic leukemia. The next decade revealed vinca alkaloids to be potential anti-cancer agents which were introduced to treat Hodgkin's disease and leukemia in the 1960s. In parallel researchers identified Camptothecin and Podophyllotoxin derivatives as potent antineoplastic agents, though it took some time to characterize their mode of action as topoisomerase inhibitors. Similarly, anthracyclines and bleomycins were first isolated in the 1950 and 1960s. Researchers began to explore different combinations of these chemotherapeutic drugs in the 1970s.

The discovery of nitrogen mustard as an anti-cancer agent was the beginning of modern cancer chemotherapy (Chabner and Roberts 2005). Ever since, more than a hundred, chemically very diverse drugs have been approved to treat cancer. They have different targets, different side effects and affect cells at different phases of the cell cycle. Many of these agents remain the backbone of current anti-cancer therapy, but they are limited by significant toxicities, a narrow therapeutic index, and resistance. All current chemotherapeutic drugs can be classified into several categories according to their mechanism of action:

(a) DNA-modifying agents
(b) Anti-metabolites
(c) Spindle poisons
(d) Topoisomerase inhibitors
(e) Cytotoxic antibiotics

Most chemotherapeutic drugs are cell cycle-specific and act on cells undergoing division. Cell cycle-specific drugs can be subdivided into S-phase- G1-phase-, G2 phase- and M-phase-specific agents according to the phase of the cell cycle in which they are active. Cell cycle-specific drugs are most effective for high growth fraction malignancies (e.g.: hematologic cancers). Their capability to kill cells displays a dose-related plateau and does not increase with further increased dosage, because at a certain time point only a subset of cells is fully drug sensitive. In contrast, cell cycle non-specific drugs such as alkylating agents have a linear dose-response curve and affect cells regardless whether they are proliferating or resting. They are effective for both low and high growth fraction tumors.

Alkylating Agents

Alkylating agents are a group of structurally diverse and highly reactive chemical compounds which represent the most widely used chemotherapeutic drugs in anti-cancer therapy (Fu et al. 2012). Alkylating agents are capable of forming covalent linkages with amino, carboxyl, sulfhydryl, and phosphate groups in biologically important macromolecules such as DNA, RNA, and proteins. The electron-rich nitrogen at the 7 position of guanine in DNA is a strong nucleophilic centre and therefore very susceptible to alkylation (Fig. 2.4). We distinguish between mono and bifunctional alkylating agents. The latter have two alkylating groups and hence, can form covalent bonds at two nucleophilic sites leading to interstrand or intrastrand cross-links (Fig. 2.4).

Interstrand cross-links are formed when the alkylating groups covalently bind to DNA bases on two opposite strands, whereas intrastrand cross-links connect DNA bases on same strand via alkylating groups. Conversely, monofunctional alkylating agents that contain only one alkylating groups fail to form interstrand or intrastrand cross-links. Monofunctional alkylating agents have less effective antitumor activity than bifunctional agents. Interestingly, more than two alkylating groups do not further increase the cytotoxic effect of these agents suggesting cross-linking of DNA to be the major cytotoxic mechanism of action of bifunctional compounds. Mismatch DNA repair activity has been shown to be a mediator of the cytotoxicity of monofunctional alkylating agents. The platinum-containing drugs don't contain alkyl groups but can form covalent links between adenine and/or guanine bases which are very similar to the ones formed by alkylating agents. Therefore, platinum-containing

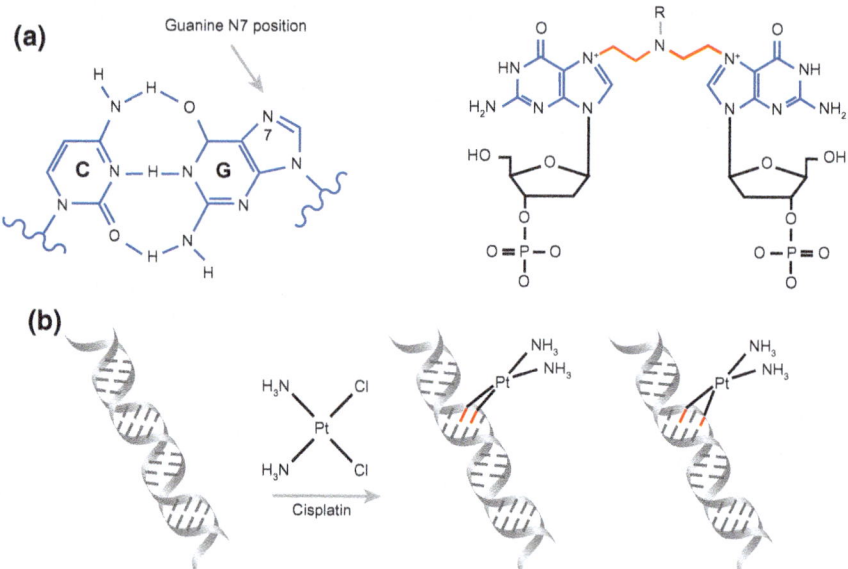

Fig. 2.4 Mode of action of alkylating agents and alkylating-like agents. **a** Alkylating agents attach an alkyl group to DNA, in particular to the guanine base of the DNA molecule. They are used in cancer treatment to damage DNA in rapidly dividing cells such as cancer cells that do not have time for DNA repair. **b** Alkylating-like agents like cisplatin can form covalent links between adenine and/or guanine bases which are very similar to the ones formed by alkylating agents but lack alkyl groups

drugs such as cisplatin or carboplatin are called **alkylating-like agents** and bind DNA bases adenine and/or guanine bases via the platinum atom. Both, classical alkylating agents and alkylating-like agents form DNA adducts that interfere with unwinding of DNA during replication. Changes in the DNA structure induced by alkylating agents or alkylating-like agents lead to cell cycle arrest preventing the synthesis of new DNA on the damaged template which allows the cell to repair the damage or leading to apoptosis if the damage is severe and cannot be repaired. These processes prevent damaged DNA from being passed on to daughter cells. Alkylating agents work in all phases of the cell cycle but their efficacy depends on cell proliferation. Therefore, their efficacy to kill cancer cells is strictly dose-dependent. Alkylating agents have still a wide range of indications including lung, breast, ovary cancers, leukemias, lymphomas, multiple myelomas and sarcomas. Several molecular mechanisms are known to be capable of conferring resistance to alkylating agents. These mechanisms can be classified in pre and post-target mechanisms. Pre-target resistance to alkylating agents prevent them from forming DNA adducts through reducing drug concentration or efficient detoxification by glutathione conjugation. In contrast, post-target mechanisms enhance the capacity of the cell to repair or tolerate the damage. There are several different classes of alkylating agents classified according to their chemical structures and mechanisms of binding including nitrogen mustards, aziridines, alkyl sulphonates, the nitrosoureas, the triazenes and hydrazines as well as the platinum-containing alkylating-like agents (Table 2.1).

The most frequently used alkylating agents in cancer therapy are nitrogen mustards, namely cyclophosphamide, isosfamide, mechlorethamine, chlorambucil and melphalan. Aziridines e.g. the currently used drugs thiotepa, mitomycin C and diaziquone are chemical compound which are closely related to the nitrogen mustards, but are less reactive. Busulfan is an alkyl sulphonate that has been used as first line treatment of chronic myelogenous leukemia (CML), before hydroxyurea and more recently imatinib were introduced though it is still in use as a component of ablative regimens prior to bone marrow transplantation. Nitrosoureas such as carmustine and lomustine are lipophilic compounds that contain a nitroso (R-NO) group and a urea. As nitrosoureas cross the blood–brain barrier, they can be used to treat brain tumors.

Table 2.1 List of alkylating agents

Drug	Drug type	Mode of action	Approval
Mechlorethamine	Synthetic	Alkylation of DNA	1949
Cyclophosphamide	Synthetic	Alkylation of DNA	1959
Carmustine	Synthetic	Alkylation of DNA	1977
Lomustine	Synthetic	Alkylation of DNA	1977
Dacarbazine	Synthetic	Alkylation of DNA	1975
Temozolomide	Synthetic	Alkylation of DNA	1999
Cisplatin	Synthetic	DNA adduct formation	1978
Oxaliplatin	Synthetic	DNA adduct formation	1996
Carboplatin	Synthetic	DNA adduct formation	1986

Triazenes and hydrazines are nitrogen-containing compounds that are metabolized to form reactive intermediates. Dacarbazine, used to treat metastatic melanoma and temozolomide a dacarbazine analogue capable of crossing the blood brain barrier an approved as a first line treatment for glioblastoma multiforme are monofunctional alkylating agents. They are unable to form intra or interstrand cross-links and rather produce methylation of DNA, predominantly on the O-6 and N-7 positions of guanylic acid. The alkylating-like agents cisplatin, carboplatin and oxaliplatin are coordination complexes of platinum and used to treat a broad range of human cancers (Box 3) (Reedijk 1999).

Box 3
The Cisplatin story
Cisplatin became a very famous drug, because it revolutionized the treatment of testicular cancer and it is an important component of combinatory treatments for several other cancer entities including colon and ovarian cancer. Nowadays, platinum-based agents are used in about 40% of all chemotherapy treatments. But the history of the discovery and development of this drug is less known. Cisplatin was synthesized in 1844 by the Italian chemist Michele Peyrone and called Peyrone's chloride. More then hundred years later in 1965 the American chemist Barnett Rosenberg discovered that bacteria extended 300 times their normal size but did not divided when placed into an electric field. However, subsequent research revealed that is was not the electrical field that inhibited cell division. Rosenberg and his colleagues figured out that the platinum electrodes that they used to generate the electrical field corroded in the ammonium chloride containing solution, producing a platinum compound. It took another effort to find out that only the cis, but not the trans-isomer conferred the effect on cell division. Cisplatin was successfully tested in animal models and entered clinical evaluation in 1972. A breakthrough pilot study evaluated Cisplatin in 15 patients with testicular cancer. 13 responded to the treatment, 7 of them completely. But there are also problems associated with the use of this drug. Cisplatin can cause severe side effects including kidney toxicity, nausea and vomiting and tumors can develop resistance to it. Nowadays nausea and vomiting can be controlled pharmacologically, and hydration can reduce kidney toxicity. The use of more focussed drug delivery strategies might further reduce the side effects. Furthermore, the development of less toxic platinum based drugs e.g. carboplatin provided alternative treatment options based on platinum drugs.

Antimetabolites
Antimetabolites are small chemical compounds that mimic endogenous molecules namely folic acids, pyrimidines or purines substituting for the normal building blocks of RNA and DNA (Shewach and Kuchta 2009). Accordingly, they are incorporated

2.3 Chemical Treatment

Table 2.2 List of antimetabolites

Drug	Drug type	Mode of action	Approval
Methotrexate	Synthetic	DHFR inhibitor	1953
Pemetrexed	Synthetic	DHFR, TS, GARFT inhibitor	2004
Fluorouracil	Synthetic	TS inhibitor	1962
Cytarabine	Synthetic	Inhibiting DNA polymerase	1969
Gemcitabine	Synthetic	Incorporated into DNA	1995
6-Mercaptpurine	Synthetic	Incorporated into DNA	1953
Fludarabine	Synthetic	Inhibiting DNA polymerase and RR	1991

DHFR dihydrofolate reductase, *TS* thymidylate synthase, *GARTF* glycinamide ribonucleotide formyltransferase, *RR* ribonucleotide reductase

into RNA and DNA or target enzymes involved in metabolic processes that give rise to the synthesis of purines and pyrimidines and are most active during the S phase of cell cycle during which DNA synthesis occurs. Antimetabolite drugs have limited effects on cells in G0 being most effective in rapidly dividing tumors. Accordingly, toxicity associated with the treatment with antimetabolite drugs is due to damage of tissues that experiences a high rate of DNA synthesis, including gastrointestinal tract and bone marrow. Antimetabolites are the earliest rationally designed anticancer drugs and are often inactive prodrugs that need to be metabolized within the body into a pharmacologically active drug. Antimetabolites are used as therapeutic agents to treat a broad range of human diseases. As antimetabolite drugs inhibit the proliferation of various immune cells they can act as immunosuppressive agents and are used for the treatment of several chronic autoimmune diseases such as rheumatoid arthritis, psoriasis and Crohn's disease. Antimetabolites such as the acyclic purine analogue acyclovir act as inhibitors of viral DNA polymerase and are efficient anti-herpes drugs, while antimetabolite inhibitors of the reverse transcriptase of HIV such as zidovudine are key components of successful AIDS treatments. Antimetabolites have been found to be useful as anti-cancer agents in 1947 when Sidney Farber observed that aminopterin was capable of inducing remissions among children with leukemia. Ever since, antimetabolites have been a rich source of efficient treatments against many types of tumors. Antimetabolites can be classified according to their structure into folic acid, pyrimidine or purine analogues (Table 2.2).

Folic acid analogues also called antifolates are antimetabolite drugs that antagonize the actions of folic acid, a necessary element for the production of nucleotides (Kamen 1997). Folic acid is the synthetic form of folate, a water-soluble vitamin (vitamin B9) found in the diet which received its name when it was first isolated from spinach leaves (Lat. follium). Folate and folic acid have no biological activity themselves, but are converted into the metabolically active tetrahydrofolate through two successive reactions, both catalyzed by dihydrofolate reductase (DHFR). Intracellular folate is converted to dihydrofolate and then reduced to tetrahydrofolate which is an important one-carbon donor in the synthesis of purines and pyrimidines. Many structural analogues of folic acid are inhibitors of DHFR, but other key enzymes in folate metabolism have also been targeted to achieve the depletion of building blocks required for the synthesis of DNA. The most commonly used folic acid analogues is **methotrexate**, one of the oldest chemotherapeutic drugs used since the early 1950s. Methotrexate is a very potent, competitive inhibitor of DHFR which selectively affects the most rapidly dividing cells (Khan et al. 2012). Methotrexate has been approved for the treatment of a range of different tumors including osteosarcoma, lung cancer, breast cancer, leukemia and lymphoma. **Pemetrexed** is a multitargeted folic acid analogue that apart from DHFR inhibits two more enzymes are involved in folate metabolism, namely thymidylate synthase, and glycinamide ribonucleotide formyltransferase. Pemetrexed is approved in combination with cisplatin for the therapy of malignant mesothelioma and as a second-line treatment of non-small-cell lung cancer (Paz-Ares et al. 2003).

Pyrimidine analogues are antimetabolites which mimic the structure of hydrophilic, six-membered heterocycles of pyrimidines. Nucleic acids contain the pyrimidine derivatives cytosine, thymine and uracil. Most pyrimidine analogues enter the cell via specialized membrane transporters as biologically inactive prodrugs and are converted into active metabolites by intracellular enzymes. The activated pyrimidine analogues can cause chain termination and inhibition of DNA synthesis when they get incorporated into DNA or inhibit enzymes required for nucleic acid production including DNA polymerases and ribonucleotide reductase. Pyrimidine animetabolite drugs used in anti-cancer therapy are analogues uracil including fluorouracil, floxuridine or capecitabine or analogues of cytosine such as azacitidine, decitabine, cytarabine and gemcitabine. **5-Fluorouracil** is one of the first examples of a drug designed based on a biochemical rational. The molecule was synthesized in 1956 to inhibit thymidylate synthase, a key enzyme involved in thymidine biosynthesis by competitive binding. 5-Fluorouracil is a prodrug structurally similar to uracil which is metabolized in the body into two active compounds that inhibit the activity of the enzyme that converts uracil to thymidine by adding a methyl group (Longley et al. 2003). While thymidylate synthase is thought to be the most important target, alternative mechanisms of 5-Fluorouracil-mediated cytotoxicity have been reported including RNA and DNA incorporation and DNA damage associated with excision repair. 5-Fluorouracil is used to treat a variety of tumor types such as colorectal, anal, oesophageal, stomach, breast, pancreatic and skin cancers. **Capecitabine** has been approved as an oral prodrug of 5-Fluorouracil. **Cytarabine** also called cytosine arabinoside is a combination of a cytosine base and an arabinose sugar used to treat hematologic neoplasms. While other antimetabolite drugs alter the base of nucleosides, cytarabine modifies the sugar component but is still capable of incorporating into RNA and DNA. Cytarabine is thought to exert its antineoplastic effect by causing damage to RNA and DNA and by inhibiting DNA polymerase. **Gemcitabine** is a deoxycytidine analogue with two fluorine atoms which is transported into the cell and converted into an active metabolite by phosphorylation (de Sousa Cavalcante and Monteiro 2014). Gemcitabine was developed in the 1980s as an antiviral agent and competes with dCTP for incorporation into DNA leading to a masked termination of DNA synthesis. Proofreading enzymes are unable to excise gemcitabine from DNA and DNA polymerases are unable to proceed. Gemcitabine has been approved for the treatment of several solid tumors. It is used as a first-line treatment for pancreatic cancer. **Azacitidine** is an analogue of cytidine known to inhibit DNA methyltransferase leading to DNA hypomethylation and in turn to increased gene expression. Therefore, azacytidine might be useful to treat tumors in which the expression of tumor suppressor genes is silenced by hypermethylation. Decitabine (5-aza-2′-deoxycytidine) is the deoxy derivative of azacytidine. Azacytidine and decitabine can also be incorporated into RNA and DNA. Both agents are used to treat myelodysplasia.

Purine analogues are antimetabolites that mimic the heterocyclic aromatic structure of purines which consists of a pyrimidine ring fused to an imidazole ring. The purine bases adenine and guanine are building blocks of DNA and RNA nucleotides. Similar to pyrimidine animetabolite drugs, most purine analogues are prodrugs that are converted intracellularly into phosphorylated derivatives. These activated purine analogues are substrates for enzymes involved in DNA synthesis and repair. Purine antimetabolite drugs exert their antineoplastic effect through depletion of purine bases, incorporation into DNA and inhibition of DNA repair. **Mercaptopurine**, a structural analogue of guanine in which an OH was replaced with a thiol group was the first purine antimetabolite to be approved for clinical applications. The intracellularly produced active metabolites of mercaptopurine inhibit the formation of purine nucleotides and interfere with DNA synthesis. The mercaptopurine metabolite deoxythioguanosine is incorporated into DNA which leads to the disruption of DNA replication. Mercaptopurine is still used for the treatment of acute lymphocytic leukemia. **Fludarabine** is a prodrug that is intracellularly converted into an active metabolite that interfere with DNA synthesis by inhibiting DNA polymerase, ribonucleotide reductase and DNA primase. Fludarabine has been approved as a second line therapy for adult patients with chronic lymphocytic leukemia (CLL).

Spindle Poisons
Spindle poisons are chemical compounds which interfere with microtubules that form the mitotic spindle required for precise division of chromosomes to produce two equal daughter cells (Matson and Stukenberg 2011). As accurate transmission of chromosomes during cell division is of foremost importance to avoid irregular chromosome content (aneuploidy) and genetic instability, cellular control mechanisms have evolved to detect errors. These regulatory mechanisms called cell cycle checkpoints ensure the accuracy of mitosis. Spindle poisons affect mitosis and engage the spindle assembly checkpoint eventually leading to cell cycle arrest and cell death. Microtubules are major components of the cytoskeleton in eukaryotic cells. They form the mitotic spindle responsible for the alignment of replicated chromosomes at the equatorial plane and their segregation to the two daughter cells. Microtubules are long, hollow cylinders made up of polymerised α- and β-tubulin dimers. α/β-tubulin dimers polymerize end-to-end by reversible non-covalent association into linear protofilaments that associate laterally to form a single microtubule (Fig. 2.5).

In most eukaryotic cells a microtubule is formed by thirteen protofilaments and has a diameter of 25 nm. The process of tubulin polymerisation by tandems results in a protein polymer with two different ends, the minus end with an exposed α-tubulin and

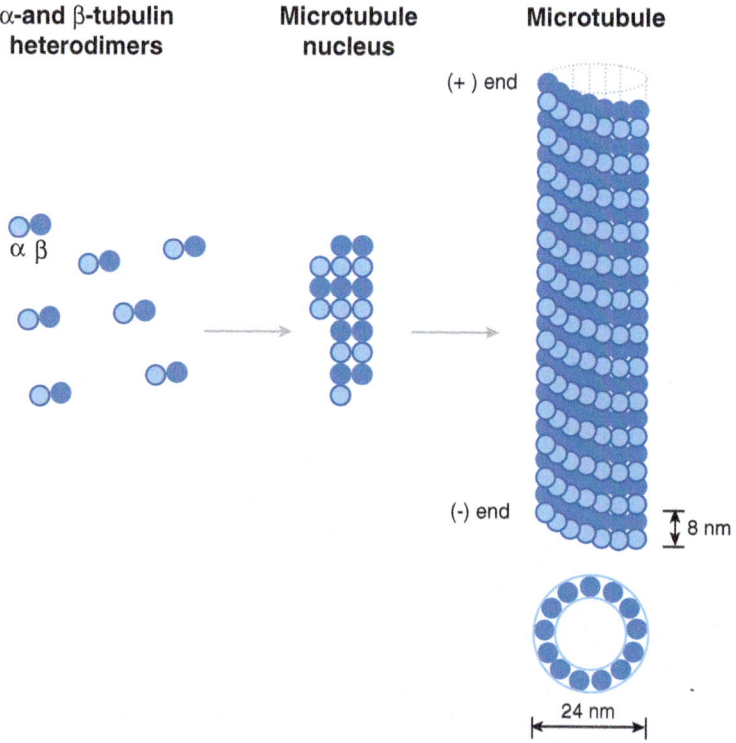

Fig. 2.5 Microtubuls dynamics. Microtubules (MTs) are dynamic structures and can assembly and disassemble. They are hollow cylindrical polymers of 24 nm diameter. MTs are made of tubulin heterodimers (α tubulin and β tubulin). MTs have two ends, a plus end (+) and a minus end (−). At the (+) end, dimers with bound GTP are assembled and GTP bound to β tubulin gets hydrolyzed to GDP. At the (−) end, dimers get dissociated and polymerization at this end is blocked

the plus end where the β-tubulin is exposed. Microtubules are very dynamic protein polymers that can polymerize and depolymerize. Microtubule dynamics are regulated by the binding and hydrolysis of GTP. α- and β-tubulin monomers have a binding site for one molecule of GTP, but only GTP bound to β-tubulin can be hydrolysed to GDP and GDP in β-tubulin is able to exchange with GTP within the soluble, unpolymerized tubulin dimer, but not in the polymerized tubulin. GTP-tubulin promotes polymerization whereas GDP-tubulin favours depolymerization. If GTP hydrolysis proceeds faster than α/β-tubulin dimers addition, the microtubule shrinks. Binding sites for vinca alcaloids, taxanes and colchicine have been described within the β-tubulin protein. The vinca binding site resides adjacent to the exchangeable GTP binding site, the taxane binding site is located in a hydrophobic pocket in the lumen of the microtubule and the colchicine binding site at the interface between α and β-tubulin. Chemical compounds targeting microtubules can be classified into microtubule-stabilizing and microtubule-destabilizing compounds (Fig. 2.6), both capable of interfering with microtubule dynamics and blocking cell division (Table 2.3). Colchicine,

2.3 Chemical Treatment

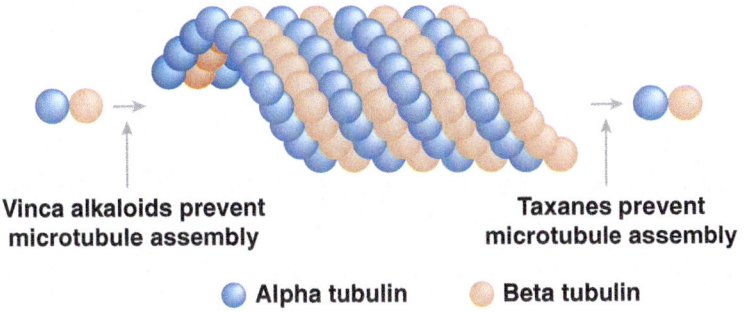

Fig. 2.6 Mode of action of taxanes and vinca alkaloids. Vinca alkaloids bind to α/β-tubulin dimers occupying tubulin's building block structure, preventing cancer cells from successfully dividing. Taxanes block dynamic instability by stabilizing GDP-bound tubulin in the microtubule

Table 2.3 List of spindle poisons

Drug	Drug type	Mode of action	Approval
Vinblastine	Natural product	Prevent microtubule assembly	1961
Vincristine	Natural product	Prevent microtubule assembly	1963
Vinorelbine	Semi-synthetic	Prevent microtubule assembly	1994
Paclitaxel	Natural product	Prevent microtubule disassembly	1993
Docetaxel	Natural product	Prevent microtubule disassembly	1995

nocodazole and vinca alkaloids inhibit the polymerization of tubulin into microtubules, while the taxanes protect microtubules against depolymerization.

The first **vinca alkaloids** vinblastine and vincristine were extracted in the 1950s from the leaves of the periwinkle plant *Catharanthus roseus* found in Madagascar. Efforts to extract and explore the therapeutic potential of compounds from this pantropical plant were prompted by the use of crude periwinkle extracts as an oral antidiabetic agent in indigenous medicine. Animal studies, however, revealed that vinca alkaloids failed to affect the blood sugar level and rather caused bone marrow suppression. Ever since naturally occurring vinca alkaloids have been used for anticancer therapy and several semi-synthetic derivatives have been developed including vinorelbine, vindesine, and vinflunine. Nowadays, vinca alkaloids are produced synthetically. Vinblastine, vincristine and vinorelbine have been approved for clinical use. As their mode of action and mechanisms of resistance do not overlap with other chemotherapeutic drugs, vinca alkaloids have been extensively used in given in combination with other drugs. They are cell cycle specific chemotherapeutic agents. They inhibit spindle formation and alignment of chromosomes and work in the M-phase of the cell cycle. **Vinblastine** like other vinca alkaloids is a dimeric structure composed of two multiringed units that reversibly binds unpolymerized tubulin dimers and polymerized tubulin at the vinca binding site in β-tubulin disrupting microtubule assembly (Silvestri 2013). Low concentrations of vinblastine preferentially affect high-affinity vinca binding sites at the end of the microtubule and decrease the rates of growth and shortening at the plus end of the microtubule. Higher drug concentrations also affect low-affinity binding sites at the microtubule surface interfering with the lateral interactions between the protofilaments, progressively exposing previously unaccessible binding sites. Accordingly, high concentrations of vinblastine result in paracrystals formed by detached spiral protofilaments, microtubule disintegration and loss of microtubule polymer mass. Clinically relevant concentrations of vinca alkaloids interfere with microtubule dynamics without affecting the polymer mass and result in metaphase arrest and cell death. Vinblastine is a component of several combination chemotherapy regimens to treat Hodgkin and non-Hodgkin lymphomas as well as testicular cancer, although it has been replaced by etoposide for the latter indication. **Vincristine** and vinblastine are chemically almost identical being the only difference a formyl instead of a methyl group substitution on the vindoline nucleus. Though the molecular mode of action is the same as for vinblastine the indications and toxicity profile is quite different. Vincristine has been used in combination with other drugs for the treatment of lymphomas, several forms of leukemias, multiple myeloma, Wilms tumor, sarcomas, neuroblastoma, rhabdomyosarcoma and small-cell lung carcinoma.

Taxanes are naturally occurring diterpenes with microtubule-stabilizing activity. The prototypic taxane paclitaxel was originally derived from the bark of the Pacific yew tree (*Taxus brevifolia*). The extract of this scare, slow growing plant was part of a collection of thousands of plant extracts screened for their anticancer activities in an effort conducted by the National Cancer Institute. As paclitaxel supply from natural sources had important environmental implications, alternative, more sustainable resources have been explored leading to the semisynthetic preparation from an inactive precursor or by complete chemical synthesis. Taxanes are large alkaloid esters that bind to tubulin by a different mechanism than the vinca alkaloids. It has been shown that taxanes bind β-tubulin on the lumenal side of microtubules, near the lateral interface between protofilaments. Taxane binding to these sites strengthen the lateral interactions between protofilaments blocking their disassembly at both ends of the microtubules, thereby protecting microtubules against depolymerization. As dynamic instability is very important for normal microtubule function during mitotic and nonmitotic phases of the cell cycle, taxane treatment results in mitotic block at the metaphase/anaphase boundary and eventually to cell death. Taxanes are poorly soluble in water and are given as slow intravenous injections. The most widely used taxanes are paclitaxel (brand name taxol) and docetaxel (brand name taxotere). Their antitumor spectra are quite similar with activity against a broad variety of cancers, many of them refractory to other therapies. **Paclitaxel** has been used for the treatment of ovarian, breast, pancreatic, cervical and lung cancer as well as Kaposi sarcoma whereas **docetaxel** is given to treat breast, head and neck, stomach, prostate and non-small-cell lung cancer (Blagosklonny and Fojo 1999).

Topoisomerase Inhibitors
Topoisomerase inhibitors are chemical compounds that interfere with the activity of specific nuclear enzymes, the topoisomerases which play an essential role in con-

trolling the topological state of DNA (Pommier 2006). **Topoisomerases** enable the DNA to solve the topological problems posed by processes such as DNA replication, transcription, chromatide separation, recombination and chromatin remodelling (Fig. 2.7). Due to the double helix structure of DNA, the **winding problem** arises when DNA strands separate during transcription and DNA replication. The progress of the replication fork during DNA replication generates over winding ahead of a replication fork and under winding behind it. Supercoiled DNA can also lead to tensional problems that prevent the transcriptional machinery from translocation along the DNA template and producing mRNA transcripts. Topoisomerases relax the helix by introducing transient DNA breaks. Topoisomerases use a tyrosyl residue as the nucleophile to attack a DNA phosphodiester bond and to establish a covalent bond to the DNA phosphate resulting in single- or double-strand breaks.

The covalent complexes of topoisomerase with cleaved DNA are known as cleavable complexes. Topoisomerases catalyze the breaking and rejoining of the DNA phosphodiester backbone and can divided into two classes, type I and type II topoisomerases. Eukaryotic **type I topoisomerases** are monomeric enzymes which cut one of the two strands of double-stranded DNA, relax the strand, and reanneal the strand. **Type II topoisomerases** are ATP-dependent heterodimeric enzymes which cut both strands of the DNA helix simultaneously to manage DNA tangles and supercoils. The level of topoisomerases is often higher in actively dividing cells including cancer cells and pharmaceutical inhibition of topoisomerases leads to growth inhibition and apoptosis. The increased requirement of topoisomerase activity in cancer cells is the rational to use topoisomerase inhibitors as anti-cancer drugs. **Topoisomerase inhibitors** can be classified into topoisomerase I and topoisomerase II inhibitors and can act as **topoisomerase poisons** and catalytic topoisomerase inhibitors (Table 2.4). **Catalytic topoisomerase inhibitors** target the catalytic activity of the enzymes and thereby prevent them from generating DNA brakes. Conversely, topoisomerase poisons stabilize the covalent complex between the enzyme and the DNA interfering with the religation step of the reaction and hence leave the DNA strands unligated.

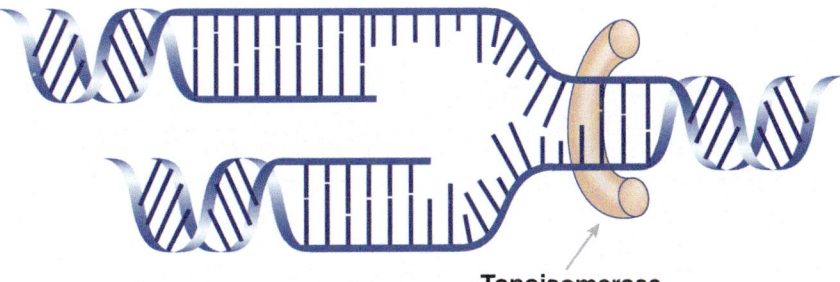

Fig. 2.7 Topoisomerases help to relieve the tension generated by unwinding DNA. DNA transcription and replication require the separation of the two strands of the DNA molecule potentially leading to overcoiling of the DNA. Topoisomerases are reversible nucleases capable of cutting the DNA backbone and resealing the phosphodiester bonds after twisting the cut ends past each other to a more relaxed configuration

Table 2.4 List of topoisomerase inhibitors

Drug	Drug type	Mode of action	Approval
Irinotecan	Synthetic	Topoisomerase I inhibitor	1996
Topotecan	Synthetic	Topoisomerase I inhibitor	2007
Etoposide	Semi-synthetic	Topoisomerase II poison	1983
Teniposide	Semi-synthetic	Topoisomerase II poison	1992
Mitoxantrone	Synthetic	Topoisomerase II poison	2000

These unrepaired DNA brakes eventually result in cell death. Topoisomerase poisoning as a mechanism of drug action emerged when topoisomerase I was identified as the molecular target of camptothecin. **Camptothecin** is an alkaloid found in the Chinese tree, Camptotheca acuminate and identified as a potent anti-cancer agent in a screening of a thousand plant extracts in the early 1960s at the US National Cancer Institute (Li et al. 2017). About 20 years later, camptothecin was found to trap the covalent DNA topoisomerase complex preventing DNA replication. Accordingly, yeast cells lacking topoisomerase I are immune to camptothecin. Further development of camptothecin was challenged by its poor water solubility, low response rates and high toxicity. Medicinal chemists generated water soluble semisynthetic camptothecin analogues with improved properties. Two of these derivatives, namely irinotecan and topotecan have been approved by FDA for their medical use in 1996 and 2007, respectively.

Irinotecan is a prodrug, which has to be hydrolyzed by carboxyl esterase to generate its active metabolite, SN-38 and is widely used for the treatment of colorectal and esophageal cancer. **Topotecan** was the first oral topoisomerase I inhibitor and is used to treat ovarian and small cell lung cancer. Despite their potent anti-cancer activity, all camptothecin analogues suffer from dose-limiting toxicities in particular bone marrow toxicity and chemical instability leading to a short half-life. Therefore, different drug delivery strategies for camptothecin analogues and the development of non-camptothecin topomerase 1 inhibitors are being actively pursued. While types I and II topoisomerases both use a conserved tyrosine to catalyse transient DNA breaks, they share little sequence homology. Type II topoisomerases are ATPases that hydrolyse ATP as an energy source for their catalytic activity. Type II topoisomerase inhibitors are chemically very diverse and include catalytic inhibitors that block the enzyme before DNA scission, interfering with DNA binding, stabilizing noncovalent DNA topoisomerase II complexes or competing with ATP. The clinical use of catalytic topoisomerase II inhibitors is limited to Aclarubicin and MST-16 to treat hematologic malignancies. The vast majority of inhibitors of type II topoisomerase though are topoisomerase poisons. **Type II topoisomerase poisons** can be classified into intercalating and non-intercalating poisons. Type II topoisomerase intercalating poisons include anthracyclines (see below section on cytotoxic antibiotics), **Mitoxantrone**, mAMSA amonafide and ellipticine. Interestingly, DNA intercalating agents like ethidium bromide are unable to poison topoisomerases, suggesting that intercalation is not sufficient to trap these enzymes n DNA. Conversely, type II

2.3 Chemical Treatment

topoisomerase non-intercalating poisons do not strongly interact with DNA and are thought to exert their effect via heir interaction with the enzyme itself. This sub-class of type II topoisomerase inhibitors include the epipodophyllotoxins **Etoposide** and **Teniposide**.

Cytotoxic Antibiotics

Cytotoxic antibiotics are a chemically diverse group of substances with microbial origin which act as anticancer agents. While the four groups of chemotherapeutic drugs discussed so far have been classified by their specific mode of action (alkylating DNA, mimicking endogenous building blocks of RNA and DNA, interfering with microtubules or blocking topoisomerase activity) cytotoxic antibiotics exert their activity against cancer cells by different molecular mechanism, though DNA is the target of most of them. Important subgroups of cytotoxic antibiotics are the anthracyclines and the bleomycins (Table 2.5). **Anthracyclines** consist of a positively charged, planar, hydrophobic tetracycline ring that facilitates their intercalation into DNA. Anthracyclines are among the most widely used and most effective anti-cancer drugs to treat a broad variety of cancer types. However, their therapeutic window is limited by cumulative dose-related cardiotoxicity which is often irreversible and acute myelosuppression. Poisoning of topoisomerase II by stabilizing DNA topoisomerase II cleavable complex (see above) has been reported to be the primary mode of action of the anthracyclines. They also intercalate between base pairs of DNA interfering with transcription and replication. In addition, anthracyclines produce reactive oxidative species damaging essential cellular components, an activity that seems to be associated with their cardiotoxicity. The first anthracyclines were isolated from Streptomyces in Italian soil samples in the early 1950s, but **Daunorubicin** was the first anthracycline antibiotic to show activity against cancer in humans. It is used by injection to treat several types of leukemias and Kaposi's sarcoma. Daunorubicin provides also the starting material for the semi-synthetic production of other anthracyclines such as doxorubicin, idarubicin and epirubicin. **Doxorubicin** is a close analogue of daunorubicin isolated from a mutated strain of Streptomyces (Fig. 2.8). It has a wider **therapeutic window** (Box 4) than daunorubicin but is still causes cardiotoxicity (Thorn et al. 2011). Importantly, doxorubicin showed a better activity against solid tumors and its clinical use was approved in the United States in 1974. It is one of the most active drugs against breast cancer.

Table 2.5 List of approved cytotoxic antibiotics

Drug	Drug type	Mode of action	Approval
Daunorubicin	Natural product	Topoisomerase II poison	1979
Doxorubicin	Biosynthesis	Topoisomerase II poison	1974
Amrubicin	Synthetic	Topoisomerase II poison	2002
Aclarubicin	Synthetic	Catalytic topoisomerase II inhibitor	1987
Bleomycin	Biosynthesis	DNA damage	1973

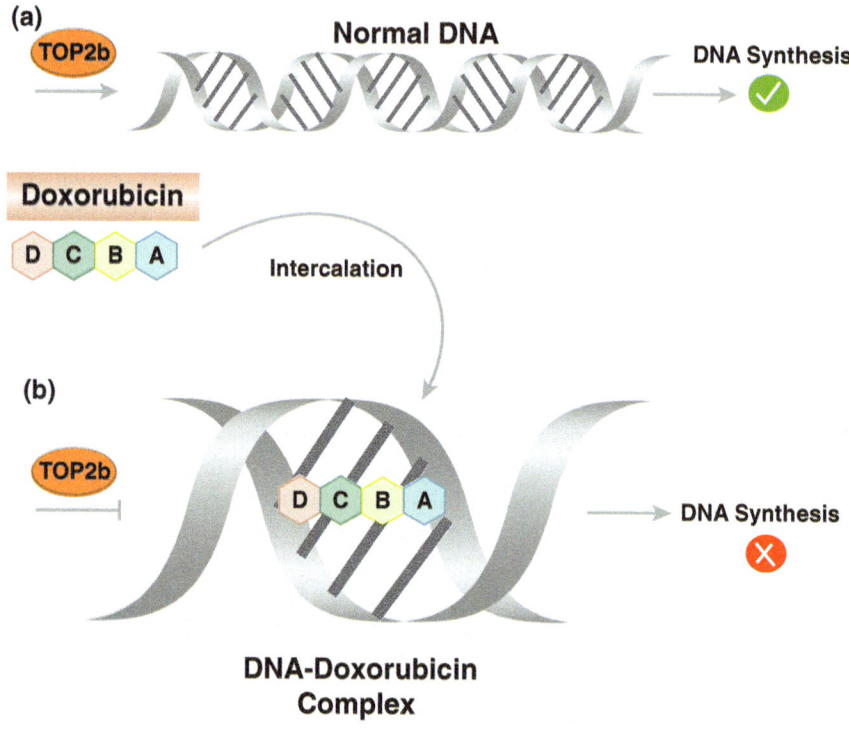

Fig. 2.8 Doxorubicin is a planar aromatic chromophore that intercalates between two base pairs of the DNA and thereby blocking the activity of Topoisomerase II. Doxorubicin is considered as a topoisomerase poison

The development of anthracyclines with better activity and less cardiotoxicity by medicinal chemistry efforts or biotechnological modification of bacterial strains is a very active research field which generated several molecules for medical use including **Amrubicin**, the first anthracycline analogue produced by de novo synthesis and approved in Japan to treat lung cancer. Similar to the anthracyclines, **bleomycins** were first isolated from Streptomyces in the 1960s while screening culture filtrates for anti-tumor activities.

Bleomycins are a family of glycopeptides that bind metal ions and oxygen to produce single-stranded and double-stranded DNA damage, similar to that generated by ionizing radiation (Chen and Stubbe 2005). Actively transcribed chromatin has been shown to be more sensitive to bleomycins. Bleomycins are large, hydrophilic molecules unable to cross cell membranes by free diffusion. Cellular uptake mechanisms might be relevant for the toxicity against specific tumor types but still remain controversial. Bleomycin was approved by FDA in 1973 and is used in combination with other drugs for the treatment of lymphomas, squamous-cell carcinomas and germ-cell tumors. Bleomycins produce low myelosuppression and low immunosuppression compared to other chemotherapeutic drugs. Nevertheless, bleomycins can cause lung inflammation which can proceed to lung fibrosis. This is the most relevant dose-limiting toxicity associated with bleomycins. Therefore, bleomycin has been used to induce pulmonary fibrosis in mouse models to investigate this condition.

> **Box 4**
> **Therapeutic window**
> The term therapeutic window also called safety window refers to a range of doses that produce a therapeutic benefit without causing unacceptable adverse effects and can be expressed as the ratio between the minimum effective dose and the minimum toxic dose. Drugs with a wider therapeutic window are safer than drugs with a low therapeutic window. For example, the anti-coagulating drug warfarin has a very small therapeutic window because the drug needs to reach a certain concentration to prevent the formation of a blood clots which can lead to heart attack, stroke, deep vein thrombosis or pulmonary embolism but a minor increase in the dose might produce severe and life-threatening bleeding.

The Limitations of Traditional Chemotherapy by Toxicity

It is very easy to kill cancer cells. Even household bleach very effectively kills cultured cancer cells. But we all know that it is not a good idea to treat patients with bleach because it would not only ablate the tumor but also kill the patient. Accordingly, the success of cancer chemotherapy is limited by problems with toxicity to healthy tissues. As most conventional chemotherapeutic agents not only affect tumor cells but also rapidly dividing cells in healthy tissues they can cause severe side effects, in particular myelosuppression, immunosuppression, alopecia, mucositis, nausea and vomiting, diarrhea and flu-like symptoms (Chabner and Roberts 2005). However, each agent has a characteristic set of toxicities, which are determined by its reactivity, metabolism, and distribution. Therefore, even structurally very similar compounds can have a very different toxicologic profile. Although some side effects are not life threatening such as nausea and vomiting which can be caused by drug effects on the central nervous system, they represent major discomfort to patients and might interfere with scheduling of therapy. Antiemetic medication such as phenothiazines, antiserotonin agents or acute doses of corticosteroids is used to help alleviate these symptoms. Dose-limiting toxicities are side effects that prevent the use of a higher drug dose which could have a therapeutic benefit. The dose-limiting toxicities of many cytotoxic chemotherapeutic drugs include myelosuppression, cardiotoxicity, nephrotoxicity, hepatotoxicity, pulmonary toxicity, dermatologic toxicity and gastrointestinal toxicity. Bone marrow suppression also called **myelosuppression** can affect different cell populations including leukocytes (white blood cells), erythrocytes (red blood cells) or thrombocytes (platelets) and therefore lead to a variety of symptoms (Fig. 2.9).

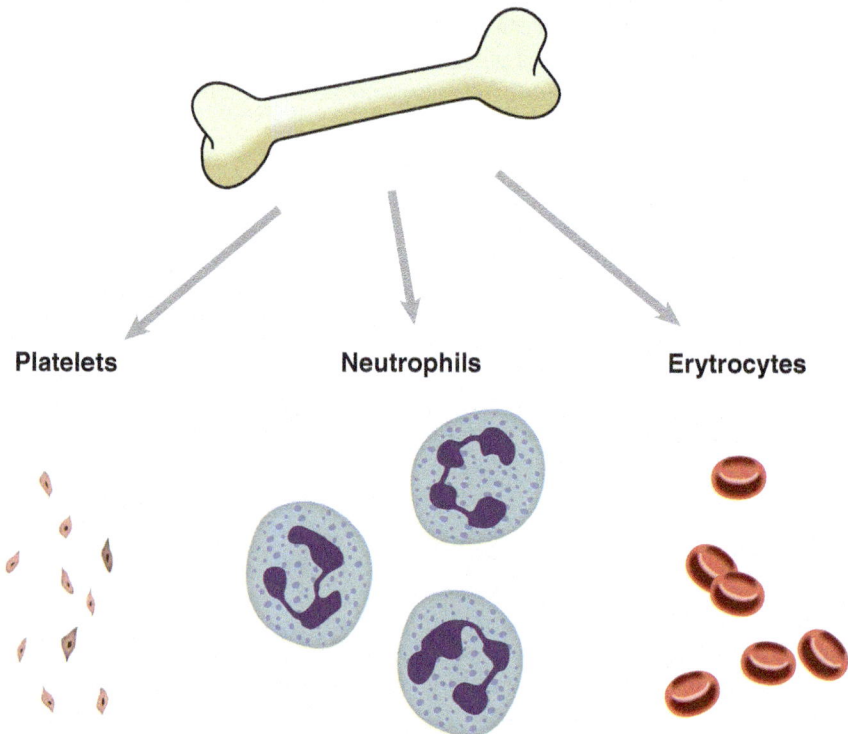

Fig. 2.9 Anemia, suppression of the immune system and decreased clotting are typical side effects of conventional chemotherapy as this type of treatment can lead to myelosuppression affecting different blood cell populations including leukocytes, erythrocytes or thrombocytes

A decrease in the production of white blood cells might compromise the immune response to invading viruses and bacteria and result in life-threatening infections. Red blood cells carry haemoglobin in their cytoplasm, a protein that binds oxygen and deliver it to all tissues in the body. Therefore, a reduced number of red blood cells leads to anaemia. Finally, as platelets play an important role in blood clotting, their decrease can result in severe bleeding. Chemotherapeutic drugs can affect the relative amount of different subtypes of blood cells. For example, taxanes and the vinca alkaloids affect mainly the generation of neutrophils, a subtype of white blood cells important to fight bacterial infections. Neutropenia, an abnormal low number of neutrophils is the principal dose-limiting toxicity of these drugs. A reduction of platelets and red blood cells is less common. Some agents with particularly high immunosuppressive activity like cyclophosphamide are used to treat autoimmune diseases. Conversely, some cytotoxic chemotherapeutic drugs including cisplatin lack a strong immunosuppressive effect or even enhance immune responses. As the immune system is essential to fight infections and its importance to eliminate tumor cells has been recognized, immunosuppression is a serious concern associated with many chemotherapeutic regimens. In order to minimize the risk of infection in the immunosuppressed patients, protective isolation measures including positive-pressure single rooms can be applied.

Anthracyclines are most notoriously associated with **cardiotoxicity**, but several alkylating agents are also known to cause damage to the heart. Myocardial damage caused by the exposure to these agents might lead to reversible cardiac dysfunction or non-reversible cardiac damage and sometimes even to life-threatening complications. Anthracyclines generate reactive oxygen species and deplete glutathione thereby compromising the key detoxification system of hydrogen peroxide in the myocardium. Cardiac function should be monitored during treatment with these agents. Dexrazoxane, a recently Food and Drug Administration (FDA)-approved drug has been reported to protect breast cancer patients against myocardial toxicity of doxorubicin. **Renal damage** has been associated with platinum-containing alkylating-like agents such as cisplatin and is usually dose-limiting. Severity of nephrotoxicity upon exposure to these agents can be reduced by high fluid intake with forced diuresis giving the osmotic diuretic Mannitol as part of the chemotherapy regimens. Since the liver is the primary site of metabolism of chemotherapeutic drugs it is not surprising that **liver toxicity** is frequently associated with the treatment with these agents. Hepatotoxicity is often based on metabolic or immune reactions to drug treatment and not always predictable. Taxanes and vinca alkaloids are also known to cause peripheral **neurotoxicity** leading to a variety of symptoms including sensory impairment, neuritic pain and motor dysfunction. The neurotoxic effect of these agents is probably due to their interference with microtubules which play an important role in axonal transport in neurons. Discontinuation or reduction of the dose or frequency of drug administration is the only means to counteract neurotoxicity. Symptoms based on drug effects in the central nervous system are less common and mainly associated with drugs that effectively cross the blood brain barrier. Many chemotherapeutic drugs including are **teratogenic**. A teratogen is defined as an agent that disturbs normal development of an embryo during pregnancy. They halt the pregnancy or can produce congenital malformation of the fetus. The toxic effect of these compounds against tumor cells and on the embryo is probably mediated through the same mechanisms. As there is a tendency of some sex specific cancers such as cervical and breast cancer to occur at earlier ages than other cancers, they might interfere with child-bearing. Drug exposure during the most sensitive period between two and eight weeks post-conception is most harmful to the embryo as it affects organogenesis. Given the mode of action of many chemotherapeutic drugs producing damage to DNA or interfering with normal cell division their **carcinogenic** potential is not surprising. Like radiation therapy, many chemotherapeutic agents can both cure and cause cancer. Those chemotherapy drugs known to cause cancer are classified as carcinogens. Chemotherapy is known as a risk factor for second cancers which occur when cancer survivors develop a new unrelated cancer. It has been estimated that about 10% of cancer survivors develop a second unrelated neoplasm caused by the treatment of the first cancer. Alkylating and alkylating-like agents as well as topoisomerase inhibitors have been shown to cause second cancers. The most frequent second neoplasm caused by chemotherapy is acute myelogenous leukemia (AML). Carcinogenic chemotherapy drugs include the nitrogen mustards, platinum based drugs and topoisomerase inhibitors. Studies have shown that the development of second cancers correlates with the dose and treatment time with alkylating agents.

The cytotoxic effect of conventional chemotherapy affects resting cells, e.g. **cancer stem cells** less effectively. Therefore, the drug might be very efficient against cells that form the bulk of the tumor, that are not able to form new cells but does not affect the rare subpopulation of cancer cells which can repopulate the tumor and cause relapse (Sagar et al. 2007). In addition, traditional chemotherapeutic agents target cell proliferation with little effect on other important hallmarks of cancers such as angiogenesis, invasion and metastases. A major problem associated with anticancer drugs (traditional and targeted therapies) is drug resistance (see Chap. 3). Some tumors, in particular pancreatic cancer, renal cell cancer, brain cancer and melanoma exhibit absence of response on the first exposure to standard agents (primary resistance). Conversely, some drug-sensitive tumors acquire resistance during the course of the treatment (acquired resistance). Drug resistance can be classified into drug-specific resistance and multi-drug resistance. Whereas drug-specific resistance is usually mediated by specific genetic alterations, the multi-drug resistant phenotype is often associated with increased expression of P-glycoprotein which expels drugs from the cell.

2.3.2 Targeted Therapy

All drugs have molecular targets however this does not mean that they are targeted therapeutics. Alkylating agents or vinca alkaloids for example target DNA or the mitotic spindle, respectively, two cellular components required for malignant cell proliferation which is the consequence of cellular transformation. Conversely, targeted therapeutics target the molecular events that drive cancer formation and progression. **Targeted therapies** are aimed at interfering with molecular events directly involved in the disease process, whereas conventional chemotherapies kill cancer cells by attacking DNA replication or cell division (Sawyers 2004). Targeted therapeutic agents interact with a specific molecular target to mediate their therapeutic effects. These molecular targets have been identified and validated through careful research as part of pathways and processes that drive tumor formation and progression. The focus of targeted drugs on specific molecular targets which are specific to a particular cancer allows for reducing collateral damage of the therapy and most approved targeted therapies have good toxicity profiles. However, as many therapeutic targets of these drugs are also present in normal cells, they display different grades of toxicity. Targeted therapies have the potential to be customized to the individual tumor profile of a patient. In order to be used in a personalized medicine setting also known as precision oncology targeted drugs have to be accompanied by companion diagnostics. **Companion diagnostics** are diagnostic tests capable of predicting if a patient is likely to benefit from treatment with a particular drug. Companion diagnostics are based on biomarkers which can stratify patiens into responders and non-responders to the therapy.

Therapeutic Targets

A **therapeutic target** is a cellular macromolecule that is involved in the pathogenesis of the disease, **druggable** (undergoes a specific interaction with a drug, see Box 5 on the concept of druggability) and its pharmacological modulation has an effect on the course of the disease (Benson et al. 2006). There are four main types of drug targets: proteins, polysaccharides, lipids, and nucleic acids. Proteins are considered the best source of drug targets as most known drugs have been shown to interact with them. The ideal therapeutic target is **specific** and **essential**. An extreme example to exemplify this concept is the bacterial cell wall as the target of Penicillin. The molecular structures of Penicillins contain a beta-lactam ring which prevents the cross-linking of the peptidoglycans of bacterial cell walls. The bacterial cell wall is essential for the bacteria because they cannot live without it and specific for bacteria, because our cell don't have one. Unfortunately, most cancers do not possess molecular targets comparable to the bacterial cell wall. Cancer research aims to identify targets that are to some degree essential and specific to cancer cells versus normal cells. It is important to note that the number of therapeutic targets is limited. Although, the human genome has been estimated to contain up to 30,000 protein-coding sequences, only a subset is disease-relevant and druggable. There might exit only be 600–1500 disease-relevant molecular targets that can be modulated by a drug. The 1578 US FDA-approved drugs act through 893 human and pathogen-derived targets (Imming et al. 2006). Among these biomolecules several specific protein families including G-coupled receptors, ion cannels and nuclear receptors dominate. The representation of kinases and phosphatases as targets for approved drugs is increasing, in particular in oncology.

Box 5
Druggability
Druggability is the likelihood of finding drugs that bind to a particular target in a disease-modifying way. The absence of cavities or pockets on the structure of the target limits its druggability. Therefore, a three-dimensional structure of the target or of a close homolog facilitates druggability assessment. Different protein families display different degrees of druggability. Many enzymes have been considered as druggable targets, whereas currently, the majority of transcription factors and adaptor proteins cannot be modulated by drugs. Protein–protein interactions which occur between big and flat surfaces and mediate many key molecular events in cancer are very attractive therapeutic targets but very challenging to modulate by small molecule drugs. Many of the most important drivers of cancer including Ras, Myc and the fusion transcription factors common in paediatric cancers are considered undruggable. It is important to note that many targets historically considered as undruggable were eventually successfully targeted. Therefore, instead of the term undruggable we should say a target is "yet to be drugged". Technological progress using sequence-targeted therapeutics such as miRNA- and siRNA-based thera-

peutics capable of raising or lowering gene expression levels at almost any point in the genome or targeted protein degradation by heterobifunctional molecules that promote degradation by bringing targets in proximity to ubiquitin ligases provides a new horizon for drugging the undruggable.

Types of Targeted Therapies

Targeted therapeutic drugs can be classified into small molecules, monoclonal antibodies, and immunotoxins. More recently, interfering RNA molecules and microRNA have been introduced. **Small molecules** are defined as molecules below a molecular weight of 900 Daltons. They rapidly diffuse across cell membranes and can reach intracellular targets as well as targets located outside the cell. Several small-molecule kinase inhibitors have been approved for clinical use. Conversely, **monoclonal antibodies** are immunoglobulin which cannot cross cell membranes and act on the outside of a cell to target specific antigens such as transmembrane receptors. They can inhibit the interaction of signaling molecules and receptors or trigger an immune response to kill cancer cells. Alternatively, immunotoxins are based on monoclonal antibodies coupled to toxic agents or radioactive molecules which can be used to guide cytotoxicity specifically to cancer cells.

Drug Nomenclature

Drugs usually have three different names, a **chemical name**, a **generic name** and a **trade name**. The chemical name is based on the chemical structure of the molecule, whereas the generic name is given during the development of the drug and used for regulatory approval procedures. Generic names use to reflect the type of drug or condition that the drug is used for. The stem "mab" at the end of a generic name, corresponds to monoclonal antibodies, the letters that precedes the "-mab" provide information concerning that the origin of the antibody. For example, "-ximab", "-zumab" and "-umab" are used for antibodies of chimeric, humanized or pure human origin, respectively. The suffix "nib" indicates small-molecule kinase inhibitors. When a drug has been approved to be introduced into the market, pharmaceutical companies sell them under a trade name. The chemical name of the tyrosine kinase inhibitor generically named imatinib is alpha-(4-methyl-1-piperazinyl)-3′-((4-(3-pyridyl)-2-pyrimidinyl) amino)-*p*-tolu-*p*-toluidide which has been sold under the trade name Gleevec. Here the generic name followed by the trade name in parentheses will be used.

Targeted Small Molecule Inhibitors

Many cell components such as polysaccharides, nuclear acids or proteins are large molecules named macromolecules that contain more than 1000 atoms. Conversely, small molecules are low molecular weight organic compounds. The upper limit of the molecular weight of small molecular compounds has been set at 900 Daltons. Their small molecular weight allows these compounds to cross cellular membranes and reach intracellular targets. Small molecule compounds can be used as research probes to characterize biological functions or to develop drugs with therapeutic activity. During the last decade many small molecule targeted drugs have been approved for their clinical use. These drugs usually interfere with the enzymatic activity of a protein. The majority of approved targeted small molecule therapeutics are kinase inhibitors. These compounds inhibit the activity of receptor tyrosine kinases, cytoplasmic protein tyrosine kinases or serine/threonine kinases. However, targeted therapy has moved beyond kinase inhibitors to include agents aimed at targeting non-kinase targets such as histone deacetylases (HDACs), Poly (ADP-Ribose)-Polymerase (PARP), the smoothened receptor (SMO) and the proteasome. It is important to note that hormonal therapy using tamoxifen or aromatase inhibitors both developed before the paradigm shift towards targeted drug discovery represents the first targeted therapy in clinical use.

Selective Estrogen Receptor Modulators and Aromatase Inhibitors
Tamoxifen (Nolvadex) is a nonsteroidal agent which competes with estrogen for binding sites of the estrogen receptor-α (ERα) inducing a conformational change and in turn inhibiting ERα-dependent gene transcription (Jordan 2003). Tamoxifen was first synthesized in 1962 and has been used as an endocrine therapy for breast cancer for the last 40 years. Almost 70% of all breast cancers express ERα which plays a major role in their tumorigenesis. Estrogen binds to intracellular estrogen receptors which upon activation translocate to the cell nucleus and regulate the expression of genes including cyclin D1, Myc, Bcl-2 and VEGF. Large patient studies revealed that ERα expression strongly correlates with the clinical response to tamoxifen. Therefore, tamoxifen is prescribed in a personalized setting, after confirmation that the tumor expresses ERα by a companion diagnostic test based on immunohistochemistry. **Aromatase inhibitors** have emerged as alternative endocrine drugs to treat breast tumors that are classified as ERα positive. Aromatase is an enzyme which catalyzes the conversion of androgens into estrogens. Therefore, aromatase inhibitors decrease the production of estrogen in peripheral tissues of the body. However, in premenopausal woman in whom most of the estrogen is produced in the ovaries, the aromatase inhibitor-induced decrease of estrogen is compensated via the hypothalamus and pituitary axis to increase gonadotropin stimulating the production of estrogen in the ovaries. Accordingly, aromatase inhibitors are used to treat postmenopausal patients with breast cancer in whom estrogen is mainly produced in peripheral tissues.

Kinase Inhibitors
Kinases are one of the largest enzyme families whose three-dimensional structures are conserved from prokaryotes to humans. Kinases play a fundamental role in the formation and progression of human tumors and are considered as readily druggable. Therefore, these enzymes have emerged as the most important class of therapeutic targets for anti-cancer therapy (Zhang et al. 2009). A great focus has been given to develop drugs that target mutationally activated kinases or oncogenic kinase fusion proteins such as BRAF, PI3K, EGFR, c-KIT, BCR-ABL and EML4-ALK. About 40 kinase inhibitors have been approved for their clinical use to date. Kinases are enzymes that phosphorylate substrates. Kinases transfer a phosphate group from an

adenosine triphosphate (ATP) molecule to a macromolecule substrate such as proteins or lipids (Fig. 2.10).

ATP donates a phosphate and generates adenosine diphosphate (ADP), whereas the substrate gains a negatively charged phosphate group. Protein kinases add phosphate groups to specific serine, threonine or tyrosine residues in their substrates. The addition of phosphate groups or their removal by phosphatases which use water to cleave phosphoric acid monoesters represents a molecular switch that regulates the activity of many signalling molecules. Approximately 30% of human proteins contain covalently attached phosphates. The human genome encodes 518 kinases. Serine and threonine are phosphorylated by Ser/Thr kinases and tyrosine residues by Tyr kinases. Kinases have two lobes and contain several conserved domains essential for their activity including an ATP binding site and a substrate binding site (Fig. 2.11).

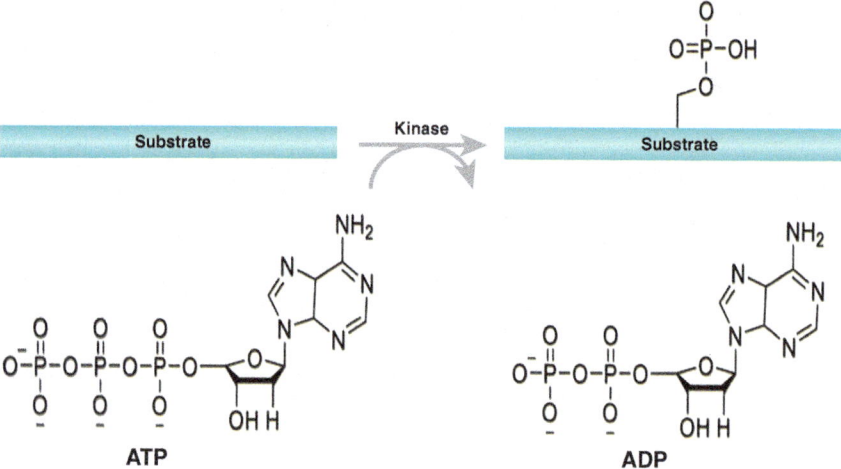

Fig. 2.10 Kinases catalyze the transfer of the γ-phosphate from ATP to a specific substrate. This reaction produces a phosphorylated substrate and ADP

Fig. 2.11 Kinases consist of two lobes (N-lobe and C-lobe) that surround the ATP-binding site. The lobes are joined by a flexible hinge region

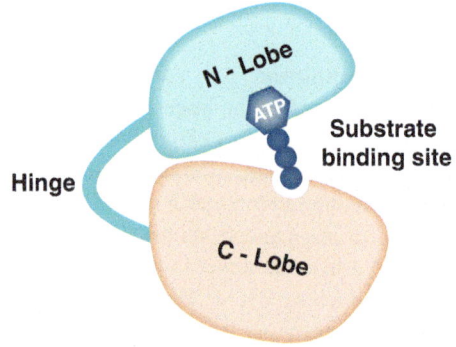

2.3 Chemical Treatment

Many kinases exist in an incompetent state and require activation before they can phosphorylate substrate. The presence of several pockets in the structure of kinases facilitated the development of kinase inhibitors capable of binding to them. Most of the kinase inhibitors compete with ATP for the binding to the ATP-binding pocket which is highly conserved among members of the kinase family. This type of compounds is called ATP-competitive inhibitors. Conversely, non-ATP competitive inhibitors bind to sites other than the ATP cleft inducing a conformational shift that prevents the kinase from phosphorylating substrates. Kinase inhibitors can also be classified into tyrosine kinase inhibitors which target receptor tyrosine kinases or non-receptor tyrosine kinases and inhibitors of serine/threonine kinases (Table 2.6).

Inhibitors of Receptor Tyrosine Kinases

Receptor tyrosine kinases (RTKs) are cell surface receptors whose extracellular domain binds to polypeptide growth factors and transduce their signals by inducing the activity of their intracellular kinase domain (Gschwind et al. 2004). RTKs comprise several families of receptors such as the epidermal growth factor receptor

Table 2.6 List of approved small molecule kinase inhibitors

Drug (Trade name)	Target	Indication
Erlotinib (Tarceva)	EGFR	NSCLC
Gefitinib (Iressa)	EGFR	NSCLC
Lapatinib (Tykerb)	EGFR, HER2	Breast Cancer
Afatinib (Gilotrif))	EGFR, HER2	NSCLC
Crizotinib (Xalkori)	ALK	NSCLC
Alectinib (Alecensa)	ALK	NSCLC
Ceritinib (Zykadia)	ALK	NSCLC
Sorafenib (Nexavar)	Multikinase	RCC, Liver cancer
Sunitinib (Sutent)	RTKs, VEGFR	RCC, GIST
Imatinib (Gleevec)	ABL, c-KIT, PDGFRα	CML, GIST, AML
Dasatinib (Sprycel)	ABL	CML, AML
Nilotinib (Tasigna)	ABL	CML
Vemurafenib (Zelboraf)	BRAF	Melanoma
Dabrafenib (Tafinlar)	BRAF	Melanoma
Trametinib (Mekinist)	MEK	Melanoma
Idelalisib (Zydelig)	PI3 Kδ	CLL
Copanlisib (Aliqopa)	PI3 Kα, PI3 Kδ	Follicular lymphoma
Temsirolimus (Torisel)	mTOR	RCC
Everolimus (Afinitor)	mTOR	RCC, Breast and pancreatic cancer

AML acute lymphocytic leukemia, *CLL* chronic lymphocytic leukemia, *CML* chronic myelogenous leukaemia, *GIST* gastrointestinal stromal tumor, *Her2* human epidermal growth factor receptor 2, *NSCLC* non–small-cell lung cancer, *RCC* renal cell carcinoma

(EGFR) family and the fibroblast growth factor receptor (FGFR) family, the vascular endothelial growth factor receptor (VEGFR) family. RTKs are activated via ligand-induced dimerization that leads to receptor auto-phosphorylation and phosphorylation of tyrosines in other proteins. As cell growth and proliferation is tightly controlled by the availability of growth factors, the overexpression or mutational activation of RTKs can promote cell proliferation independent of growth factors. Accordingly, RTKs are key drivers of the formation and progression of a wide range of cancers and the single most important class of therapeutic targets for anti-cancer therapy. The majority of the approved kinase inhibitors are RTK inhibitors including Afatinib (Gilotrif), Axitinib (Inlyta), Cabozantinib (Cabometyx), Erlotinib (Tarceva), Gefitinib (Iressa), Lapatinib (Tykerb), Lenvatinib (Lenvima), Nilotinib (Tasigna), Pazopanib (Votrient), Ponatinib (Iclusig), Sorafenib (Nexavar), Sunitinib (Sutent) and Vandetanib (Caprelsa). Most of these drugs target more than one kinase.

EGFR Inhibitors
The identification of mutations within the EGFR in patients with a range of solid cancers including non-small-cell lung cancer (NSCLC) promoted the search for inhibitors specific for EGFR. Drugs capable of blocking EGFR include monoclonal antibodies against EGFR (see below) and small molecule RTK inhibitors. Based on the screening of a library of chemical compounds, the most active inhibitors against EGFR were selected and **Gefitinib (Iressa)** was developed as the first selective kinase inhibitor of EGFR. Gefitinib was followed by **Erlotinib (Tarceva)** (Kim 2014). Gefitinib and erlotinib are both orally active ATP competitive EGFR inhibitors, but chemically unrelated and with a different toxicity profile. The path to approval for this first generation of EGFR inhibitors has been difficult as many clinical trials did not provide conclusive results. The first phase 3 clinical trials that combined gefitinib or erlotinib with chemotherapy in unselected patients with NSCLC failed to show any benefit. The clinical development of these two drugs underscores the importance of companion diagnostic tools to select patients that are predicted to respond to the treatment. Only around 10% of North American and European NSCLC patients respond to gefitinib. Patient response rates were shown to be higher in patients that had never-smoked, females, adenocarcinoma histotypes, and East Asians. Eventually, in 2004 researchers discovered that activating mutations in the kinase domain of EGFR which are present in approximately 10–15% of NSCLC patients predicted sensitivity to EGFR inhibitors. These drugs are now the first line treatment (Box 6) for NSCLC with activating EGFR mutations and the identification of these mutations is mandatory prior to patient selection. In this personalized medicine setting more than 70% of cases can be controlled until resistance is acquired (typically between 11 and 14 months post-treatment). More recently, the dual tyrosine kinase inhibitors **Afatinib (Gilotrif)** and **Lapatinib (Tykerb)** which inhibit not only the kinase activity of EGRF, but also that of the human epidermal growth factor receptor 2 (Her2) have been developed and approved for the treatment of patients with NSCLC and breast cancer respectively.

2.3 Chemical Treatment

> **Box 6**
> **First line treatment**
> First line treatment is the treatment regimen recommended to be the initial treatment for a given cancer and is based on evidence in clinical practice and presented as written guidelines or practices of the community of oncologists. These recommendations can change if new evidence becomes available. The guidelines provide the name of the drug or the drug combination, dose, timing, length of treatment, mode of administration. There is no law in place that doctors have to use the accepted first line treatments and many variations might be applied according to the age, health and comorbidities of a patient. When the first line treatment fails to meet the expectation, looses efficacy or produces unacceptable side effects, a second line treatment might be used. Many new drugs are approved as a second line treatment to treat cancers which already have established first line treatments.

ALK Inhibitors

Anaplastic lymphoma kinase (ALK) is an oncogenic RTK involved in several fusion genes. A sub-group of about 5% of NSCLC patients carry a chromosome inversion that encodes the fused oncogene EML4-ALK. This genetic alteration was discovered in 2007 and shown to increase the enzymatic activity of ALK driving tumorigenesis. At about the same time, **Crizotinib (Xalkori)**, a small molecule ALK inhibitor that was identified and developed by structure based drug design aimed at targeting the receptor for hepatocyte growth factor MET, another RTK. Crizotinib was shown to inhibit MET and ALK-mediated tyrosine phosphorylation in cell lines and to suppress tumors expressing EML4–ALK fusion proteins in xenografted mice. In contrast to gefitinib and erlotinib, crizotinib was evaluated right from the beginning in clinical trials with selected patients demonstrating ALK-positive NSCLC (Shaw et al. 2013). After only four years of clinical development, crizotinib was approved by the FDA for ALK-rearranged NSCLC with a patient response rate of almost 60%. However, this also means that over 40% of ALK-positive NSCLC patients fail to respond to the treatment with crizotinib. Worse still, the initial responders eventually develop resistance to crizotinib. Consequently, the characterization of mechanisms that can confer either intrinsic or acquired resistance in the presence of an ALK fusion protein could lead to the development of biomarkers to further stratify NSCLC patients or identify new therapeutic strategies to overcome crizotinib resistance. More recently, second generation ALK inhibitors have been approved for the treatment of patients with ALK-positive NSCLC which include the small molecule RTK inhibitors **Alectinib (Alecensa)** and **Ceritinib (Zykadia)**.

VEGFR Inhibitors

The observation that vascular endothelial growth factor (VEGF) activated VEGFR plays a critical role for the growth of blood vessels to supply the growing tumor mass with oxygen and nutrients led to the development of VEGFR inhibitors. The process of the formation of new blood vessels from pre-existing vessels is called **angiogenesis** and is known to be important for tumor progression because tumors cannot grow larger than 2 mm without angiogenesis. Many human tumors present high concentrations of the angiogenic factors VEGF. Drugs capable of blocking VEGF/VEGFR include monoclonal antibodies against VEGF or VEGFR (see below) and small molecule RTK inhibitors inhibit angiogenesis and therefore are useful to starve the tumor. **Sunitinib (Sutent)** and **Sorafenib (Nexavar)** are oral, small molecule inhibitors that inhibit the activity of a variety of kinases (multi-targeted kinase inhibitors) including VEGFR and their clinical benefit is thought to be based on their anti-angiogenic activity (Wilhelm et al. 2006). Sunitinib is approved for the treatment of renal cell carcinoma (RCC) and imatinib-resistant gastrointestinal stromal tumor (GIST). Sorafenib has been approved for the treatment of kidney and liver cancer as well as for radioactive iodine resistant advanced thyroid carcinoma.

Inhibitors of Non-receptor Tyrosine Kinases

Non-receptor tyrosine kinases are a family of about 30 cytosolic enzymes in humans and include several kinases shown to be important drivers of cancer such as Src, Fyn, Abl and Jak, Fak, Btk. As these kinases lack an extracellular domain they are mainly targeted by small molecule inhibitors which are capable of passing the cell membrane and act inside the cell. The small molecule kinase inhibitor **Imatinib (Gleevec)** emerged as a paradigm for molecularly targeted therapies. Gleevec was introduced in 2001 for the treatment of Chronic Myelogenous Leukaemia (CML) (Capdeville et al. 2002). CML is a cancer of the white blood cells caused by the reciprocal translocation between chromosome 9 and chromosome 22. The resulting Philadelphia chromosome contains the fusion of the Bcr and Abl genes that gives rise to a constitutively active kinase enzyme. BCR-ABL is a non-receptor tyrosine kinase localized in the cytoplasm or nucleus of the cell. Imatinib prevents signal transduction of BCR-ABL by binding to its ATP binding site. This prevents the transfer of phosphate groups from ATP to a protein substrate and suppresses cell growth and division (Fig. 2.12).

The success of Imatinib has proven that the concept of targeting specific molecular events in cancer can result in highly efficient anticancer therapies. Nevertheless,

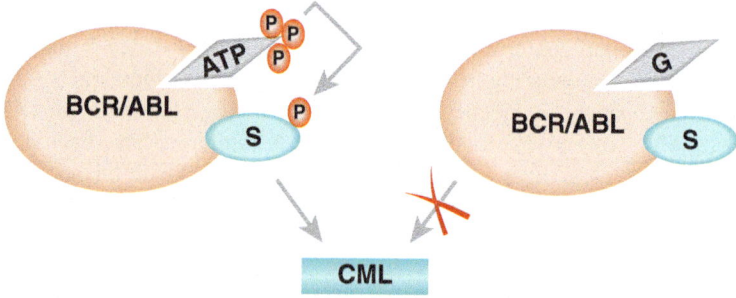

Fig. 2.12 Imatinib (G) binds to the ATP binding site of BCR-ABL and prevents the phosphorylation of substrates

as CML is a genetically simple neoplasm caused by a single aberrant protein there is still substantial debate about whether the Imatinib-paradigm can be translated to other cancers which are caused by a multitude of complex interacting genetic and environmental factors. The observation that Imatinib also inhibited the receptor tyrosine kinases KIT and PDGFRα involved in Gastrointestinal stromal tumors (GIST) provided a strong rational to test imatinib for the treatment of this disease. GIST is a good example of how the molecular understanding of a disease has resulted in a new treatment option. GISTs are mesenchymal tumors derived from the gastrointestinal tract. The pathogenesis of GISTs is driven by mutually exclusive mutations in the *mast/ stem cell growth factor receptor (SCFR)* (also known as proto-oncogene *c-KIT*) or *alpha-type platelet-derived growth factor receptor (PDGFRα)* genes that lead to the constitutive activation of these tyrosine kinases. Imatinib obtained U.S. regulatory approval for the treatment of patients with advanced or unresectable GIST and in an adjuvant setting after the surgical removal of KIT-positive tumors. As acquired resistance to Imatinib limits the efficacy of this drug in a subset of patients, second generation of BCR-ABL inhibitors have been developed among them **Dasatinib (Sprycel)** and **Nilotinib (Tasigna)** approved for the treatment of Philadelphia chromosome positive CMLs.

Inhibitors of Serine/Threonine Kinases
Serine/threonine kinases represent the biggest family of kinases and several of them have been shown to be important drivers of tumorigenesis including BRAF, MEK, PI3K, AKT and mTOR. Most serine/threonine kinases are cytoplasmic enzymes and therefore mainly targeted by small molecule inhibitors capable of passing through the cell membranes (Zhang et al. 2009). BRAF and MEK are components of the mitogen-activated protein kinase (MAPK) signalling pathway. The activation of the MAPK pathway is the result of somatic mutations within the BRAF protein, a member of the Raf kinase family present in over half of all melanomas or N-RAS, which are associated with about 15% of melanomas. The most frequent activating mutation in BRAF consists of the substitution of glutamic acid for valine at amino acid 600 (V600E). The fact that the MAPK signaling is activated in the majority of melanomas provided a strong rational for targeting this pathway with specific inhibitors. Using a fragment-based lead discovery approach to target the ATP binding pocket within the BRAF protein, several compounds were identified that specifically inhibit mutant BRAF and abolished BRAF-dependent downstream signaling. After evaluating the anti-proliferative effect in vitro and in vivo as well as pharmacokinetics properties, the compound PLX4032, (known as vemurafenib) was selected and successfully tested in clinical trials. **Vemurafenib (Zelboraf)** was approved for the treatment of BRAF-mutated metastatic melanoma in the United States in 2011 and the European Union in 2012 (Chapman et al. 2011). As vemurafenib only induces regression in tumors that express a mutant form of BRAF (BRAF-V600) companion diagnostics are essential for therapeutic success with this therapeutic. In fact, vemurafenib and testing for the presence of BRAF mutations are a model example of companion diagnostics (Box 7) associated with improved clinical response and survival in metastatic melanoma patients. A second inhibitor of BRAF-V600, **Dabrafenib**

(**Tafinlar**) with a slightly different toxicity profile was submitted to the FDA for approval in 2013 to treat BRAF-V600 positive advanced melanoma. Despite the clinical success of these BRAF inhibitors, most treated tumors develop resistance that results in disease progression within one year. As the reactivation of the MAPK pathway has been shown to be involved in the resistance to BRAF-inhibitors in many patients, a rational sequencing or combination of BRAF-inhibitors with other agents that inhibit the pathway downstream of BRAF has been pursued (Robert et al. 2015a). Recently, high-throughput screening for compounds that induce expression of p15INK4b identified **Trametinib (Mekinist)**, a very specific allosteric inhibitor of MEK1/2 that was successfully tested in pre-clinical experiments and clinical trials, receiving U.S. regulatory approval as a monotherapy for the treatment of patients with BRAF-V600E mutated metastatic melanoma in 2013. However, as in the case of the BRAF inhibitors, patients develop resistance against Trametinib within 6–7 months of treatment. To combat resistance to MAPK inhibitors, Trametinib was tested in combination with dabrafenib. This dual combination of BRAF and MEK inhibitors was the first successful attempt to combine targeted therapies to overcome acquired drug resistance in an oncogene-defined patient population (Flaherty et al. 2012) leading to its approval by FDA in 2014 (Flaherty et al. 2012).

> **Box 7**
> **Companion Diagnostics and Personalized Medicine**
> Ideally, targeted therapies go hand in hand with companion diagnostics that allows the stratification of patients into predictive responders and non-responders. Companion diagnostics are biomarkers that accompany specific therapeutic drugs to select or exclude patient groups for treatment according to biological criteria e.g. the presence of the molecular target. This patient analysis enables treatment customization to the individual characteristics of each patient. This rationally individualized approach to disease treatment has become known as personalized medicine or precision medicine. In oncology, patient stratification through companion diagnostic tools is increasingly incorporated into clinical practice. Biomarker-driven tests to screen patients for possible benefit from targeted treatment prevent non-responders from suffering side-effects from very expensive drugs without any therapeutic effect. Targeted drugs with approved companion diagnostics include Imatinib, Crizotinib, Trastuzumab, Vemurafinib, Trametinib and Olaparib.

The Phosphatidylinositide 3-kinases (PI3K)/AKT cascade is considered as the most frequently activated signalling pathway in human cancer (Hennessy et al. 2005). PI3K is a lipid kinase that phosphorylates phosphatidylinositides, producing phosphatidylinositol-3,4,5-trisphosphate (PIP3), a second messenger that recruits AKT to the cell membrane. AKT is a serine/threonine kinase and the major effector of this pathway. AKT phosphorylates a broad range of proteins which play important roles in cell physiology. One of the most important downstream components of this pathway in human malignancies is mammalian target of rapamycin (mTOR). Accordingly, much effort has been dedicated developing small molecule inhibitors of PI3K, AKT and mTOR. Currently, inhibitors approved for their clinical use include the PI3K inhibitors Idelalisib (Zydelig), Copanlisib (Aliqopa) and the mTOR inhibitors Temsirolimus (Torisel) and Everolimus (Afinitor). **Idelalisib (Zydelig)** is a specific inhibitor of the PI3 Kδ isoform which is expressed in B-cells. PI3Kδ inhibition in these cells abolishes proliferation and induces apoptosis. Accordingly, Idelalisib is used as a second-line treatment for patients with chronic lymphocytic leukemia (CLL) (Gopal et al. 2014). Recently, **Copanlisib (Aliqopa)**, an inhibitor that predominantly reduces the enzymatic activity of the PI3Kα and PI3Kδ isoforms has been approved in the USA to treat patients with relapsed follicular lymphoma (Markham 2017). mTOR is a serine/threonine kinase related to PI3K kinases which receives environmental inputs to coordinate cell metabolism and growth, two processes known to be significantly altered in cancer cells. A pharmacological inhibitor of mTOR, Rapamycin was already available before the protein was even discovered. mTOR associates with different regulatory subunits in complexes with distinct signalling functions, namely mTORC1 and mTORC2 (Sabatini 2006). Rapamycin acts as an allosteric inhibitor of mTOR and has immunosuppressive activity. Rapamycin and its analogs, called rapalogs preferentially inhibit the nutrient-sensitive signaling complex mTORC1, but not the mTORC2 complex. Two rapamycin derivatives, the rapamycin pro-drug **Temsirolimus (Torisel)** and **Everolimus (Afinitor)** have been approved for the treatment of renal cell carcinoma in 2007 and 2009, respectively. More recently, **Everolimus** has also been approved as a component of combinatory regimens to treat advanced breast cancer, pancreatic neuroendocrine tumors.

HDAC Inhibitors

Histone deacetylases (HDACs) which are comprised of four classes (HDAC I, II, III, and IV) and use zinc- or NAD+-dependent mechanisms to remove acetyl groups from histones and other proteins. Acetyl groups added by histone acetyltransferases (HATs) neutralize the positive charge of histones and diminish their affinity to negatively charged DNA. Conversely, HDACs promote high-affinity binding between histones and the DNA backbone generating condensed chromatin structures associated with the suppression of gene expression (Fig. 2.13).

In addition, HDACs deacetylate and in turn regulate the activity of non-histone proteins including signalling proteins and transcription factors. Many proteins whose expression and activity is regulated by HDACs are of key importance to tumor initiation and progression (Glozak and Seto 2007). Accordingly, small molecule inhibitors of HDACs have been developed as possible treatments for cancers (Table 2.7). HDAC inhibitors have been shown to attenuate cell proliferation, promote apoptosis and cell differentiation by molecular mechanism that are not fully understood. Several HDAC inhibitors are available for clinical use to treat cancer including Vorinostat (Zolinza), Romidepsin (Istodax), Chidamide (Epidaza), Panobinostat (Farydak), Belinostat (Beleodaq). HDAC inhibitors differ in their chemical structure and selectivity against the different HDAC classes. The first HDAC inhibitor to be approved was **Vorinostat (Zolinza)** in 2006 (Grant et al. 2007). Most HDAC inhibitors such

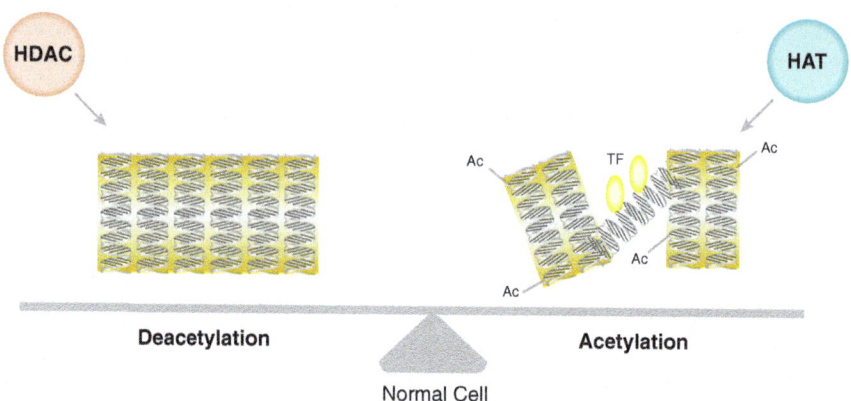

Fig. 2.13 Histone deacetylation mediated by Histone deacetylases (HDACs) prevents gene expression whereas histon acetylation mediated by histone acetyltransferases (HATs) allows gene expression. TF, Transcription factor

Table 2.7 List of approved HDAC inhibitors

Drug (Trade name)	Target	Indication
Vorinostat (Zolinza)	HDAC	CTCL
Romidepsin (Istodax)	HDAC	CTCL, PTCL
Chidamide (Epidaza)	HDAC	PTCL
Panobinostat (Farydak)	HDAC	MM
Belinostat (Beleodaq)	HDAC	PTCL

CTCL cutaneous T cell lymphoma, *PTCL* peripheral T-cell lymphoma, *MM* Multiple myeloma

as Vorinostat act as chelator for zinc ions in the active site of HDACs inhibiting their enzymatic activity and in turn increasing the amount of acetylated histones and acetylated proteins. Vorinostat and Romidepsin have been approved for the treatment of cutaneous T cell lymphoma (CTCL). **Panobinostat (Farydak)** is a non-selective HDAC inhibitor approved as part of a combinatory regimen together with bortezomib and dexamethasone for the treatment of multiple myeloma.

PARP Inhibitors
Poly ADP ribose polymerases (PARPs) are a family of enzymes that catalyze poly ADP-ribosylation, a process consisting in the transfer of ADP-ribose to target proteins (Jagtap and Szabo 2005). Most of these proteins are nuclear proteins and involved in recombination, replication, transcription and DNA repair. PARP enzymes are important components of base excision repair (BER) and nucleotide excision repair (NER) and their expression and activity have been shown to be induced upon DNA damage. Tumors with mutations in the BRCA1, BRCA2 or PALB2 gene and therefore defective in the repair of double-strand DNA by homologous recombina-

2.3 Chemical Treatment

tion repair (HRR) mechanism, rely more on PARP-mediated DNA repair and are sensitive to PARP inhibition (Fig. 2.14).

Hereditary BRCA1 or BRCA2 mutations are known to genetically predispose to develop cancer and have been identified in a subset of patients with ovarian, breast, and prostate cancers. Normal cells are less sensitive to PARP inhibitors as they replicate less, accumulate less DNA damage and their repair via HRR is intact. Loss of both, PARP and BRCA functions has disastrous consequences for the cell, presenting a classical synthetic lethal genetic interaction (Fong et al. 2009). PARP inhibitors are small molecule compounds that block their enzymatic activity of PARPs and trap them on DNA (Table 2.8). **Olaparib (Lynparza)** has been approved as a second line treatment for patients with ovarian and HER2-negative metastatic breast cancer with germline BRCA mutations. Similarly, **Rucaparib (Rubraca)** has been recently approved for pre-treated BRCA-mutant ovarian cancer and **Niraparib (Zejula)** to treat ovarian, fallopian tube, and primary peritoneal cancer. PARP inhibitors are being tested in combinatory treatment regimens e.g. together with radiation therapy with lower radiation dose and have been reported to efficient in platinum-sensitive cancers.

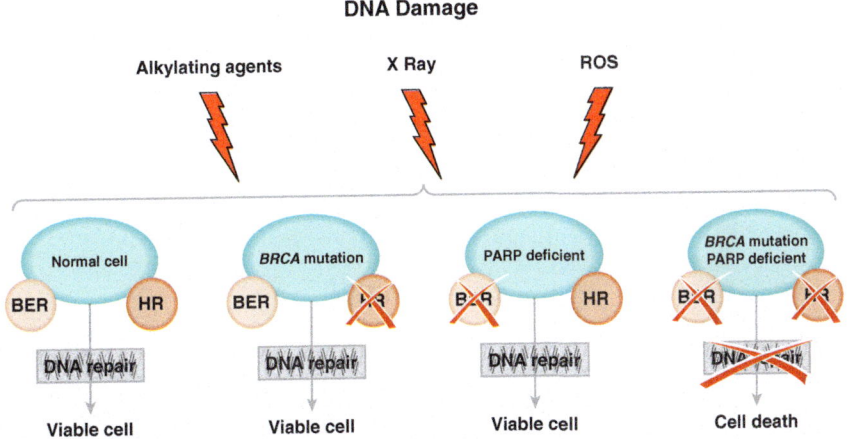

Fig. 2.14 Synthetic lethality of PARP inhibition and BRCA mutations. BER, base excision repair; HR, homologous recombination repair (HRR). PARP inhibitors effectively kill cancer cells defective in the BRCA1 or BRCA2 genes. Adapted from Oncology, 24, 1

Table 2.8 List of approved PARP inhibitors

Drug (Trade name)	Target	Indication
Olaparib (Lynparza)	PARP	Breast and ovarian cancer
Rucaparib (Rubraca)	PARP	Ovarian cancer
Niraparib (Zejula)	PARP	Ovarian cancer

SMO Inhibitors

Smoothened (SMO) is a 7-transmembrane protein which is usually inhibited by the receptor Patched1 (PTCH1) (Corbit et al. 2005). PTCH1 and SMO are the upstream elements of the sonic hedgehog (Shh) signaling pathway which is important for cell differentiation, proliferation and polarity. Similar to the Notch and Wnt signaling pathways, the Shh signalling cascade depends on regulated proteolysis of latent gene regulatory proteins. When the Shh pathway is inactive, the downstream glioma-associated (GLI) transcription factor is cleaved and Gli-dependent transcription of target genes blocked. Binding of the PTCH1 receptor by its ligand Shh causes PTCH to release SMO protein from inhibition. SMO then emits downstream signals that protect cytoplasmic Gli protein from cleavage and in turn intact Gli can then migrate to the nucleus and activate transcription of target genes (Fig. 2.15).

Hereditary mutations in the PTCH1 gene cause Gorlin syndrome, a genetic disorder associated with craniofacial and skeletal alterations and an increased risk to develop basal cell carcinoma and medulloblastoma. Spontaneous basal cell carcinomas and medulloblastomas frequently contain somatic mutations in components of the Shh pathway including PTCH1, SMO, and Suppressor of Fused (SUFU). Furthermore, Shh signalling plays a key role in the regulation and maintenance of cancer stem cells. Accordingly, therapeutic inhibition of this pathway is considered as a very attractive strategy to treat several different types of cancers, in particular tumors that result from excessive Shh signalling such as basal cell carcinoma and medulloblastoma. In the 1960s **Cyclopamine**, a steroidal alkaloid isolated from the corn lily was identified as a poison that blocks the Shh signalling pathway. Cyclopamine was shown to prevent the developing brain of lambs from separating

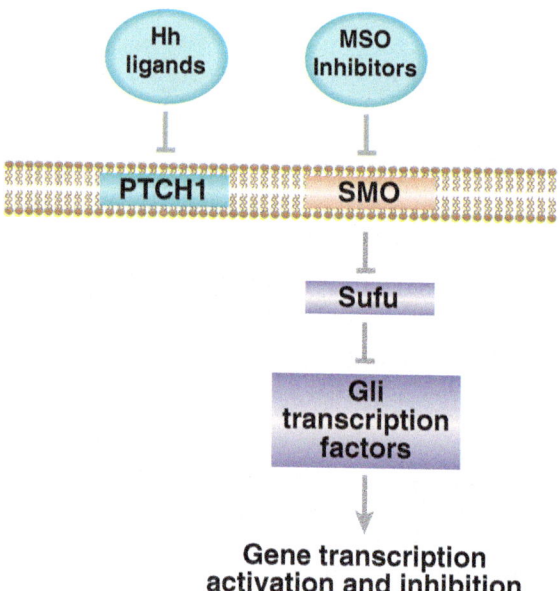

Fig. 2.15 SMO inhibitors prevent Gli mediated transcription. Smoothened (SMO) is inhibited by the transmembrane receptor Patched (PTCH1) in the absence of ligands. In the presence of ligands, the inhibition of SMO is released and SMO interacts with suppressor of fused (Sufu) promoting nuclear translocation and activation of Gli transcription factors

Table 2.9 List of approved Smoothened inhibitors

Drug (Trade name)	Target	Indication
Vismodegib (Erivedge)	Smoothened	BCC
Sonidegib (Odomzo)	Smoothened	BCC

BCC basal cell carcinoma

into two lobes resulting in an alobar brain and causing cyclopia, the development of a single eye. Cyclopamine was named after this birth defect. Cyclopamine prevents the conformational shift necessary to activate SMO by binding to its transmembrane domain. Therefore, in the presence of Cyclopamine, SMO is unable to propagate the Shh signal and the pathway is interrupted. Cyclopamine never reached clinical testing as is causes severe toxicity in mice and has poor solubility. However, it provided chemists with very useful information about the chemical structure required for its functionality. Based on the pharmacophore (see definition in the section on lead optimization) of Cyclopamine, chemical derivatives have been developed with improved efficacy and better solubility (Table 2.9) (Ruat et al. 2014). **Vismodegib (Erivedge)** was the first approved drug that targeted the Shh signalling pathway. Vismodegib is a Cyclopamine-competitive SMO antagonist approved for the treatment of basal cell carcinoma (BCC) in 2012. Three years later **Sonidegib (Odomzo)** was approved for the same indication. Vismodegib and Sonidegib are administered orally. Cyclopamine-competitive SMO inhibitors are being evaluated in numerous clinical trial for patients with different cancers including medulloblastoma, glioblastoma, small-cell lung cancer, stomach cancer, pancreatic cancer, colorectal cancer and multiple myeloma.

Proteasome Inhibitors

The proteasome is a cylindrical multi-subunit protein complex found in the nucleus and cytoplasm of eukaryotic cells which degrades obsolete or misfolded proteins (Voorhees and Orlowski 2006). The 26S proteasome (S, Svedberg unit for sedimentation rate) is composed of three subunits, a 20S catalytic core capped with two 19S regulatory subunits (Fig. 2.16).

The selective hydrolysis of client proteins by the proteasome is essential to ensure the appropriate concentration of specific proteins and the clearance of abnormal protein products. Proteins destined for proteasomal degradation are tagged by a covalently linked small protein called ubiquitin. The subsequent attachment of additional ubiquitin molecules results in a polyubiquitylated protein substrate for ATP-dependent proteolysis by the proteasome. The proteases involved in this process hydrolyze peptide bonds of the substrate and produce peptides of about seven amino acids. The ubiquitin proteasome pathway is responsible for the degradation of more than 80% of cellular proteins and essential for protein recycling, protein quality control and maintenance of protein homeostasis. Early studies revealed the antiproliferative and pro-apoptotic effect of proteasome inhibition and demonstrated that malignant cells display higher sensitivity to this treatment compared to normal cells. The mechanism responsible for this differential susceptibility remains

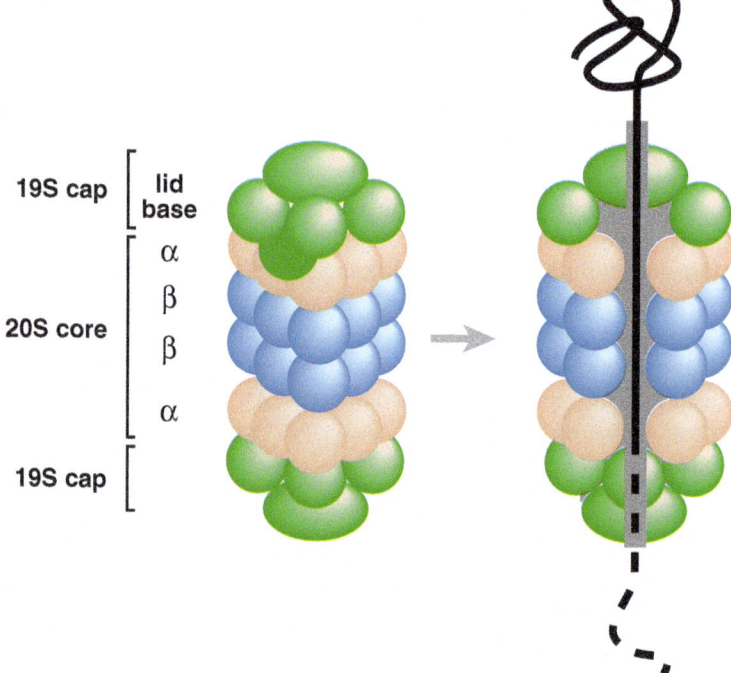

Fig. 2.16 Structure of the 26S proteasome. The 20S catalytic core capped with two 19S regulatory subunits building a 26S proteasome

to be characterized but the higher proliferative rate and increased protein synthesis of cancer cells might lead to a stronger reliance on proteasome function and make them more vulnerable to proteasome inhibition (Manasanch and Orlowski 2017). Accordingly, the rate of translation in multiple myeloma cells correlates with the sensitivity to proteasome inhibition. In that sense, proteasome inhibitors share with standard chemotherapeutic drugs the focus on consequences of cellular transformation rather than targeting molecular alteration that drive tumorigenesis and tumor progression. However, proteasome inhibitors are more efficient in specific tumors and mechanisms such as reduced degradation of p53 and inhibition of NFκB through stabilization of its inhibitor IκB have been proposed to account for this specificity (Table 2.10). **Bortezomib (Velcade)** was the first proteasome inhibitor approved for its clinical use in 2003. It was identified based on its cytotoxicity in a screening of boronic acid analogues against 60 cancer cell lines available at the US National Cancer Institute. Bortezomib is a peptide analogue that reversibly binds the catalytic site of the 26S proteasome. The preclinical and clinical development of Bortezomib revealed a particular efficacy against multiple myeloma and it was rapidly approved for the treatment of this disease. However, Bortezomib is associated with several limitations including peripheral neuropathy as the dose limiting toxicity and its requirement for intravenous administration, prompting the

2.3 Chemical Treatment

Table 2.10 List of approved Smoothened inhibitors

Drug (Trade name)	Target	Indication
Bortezomib (Velcade)	Proteasome	MM, MCL
Carfilzomib (Kyprolis)	Proteasome	MM
Ixazomib (Ninlaro)	Proteasome	MM

BCC basal cell carcinoma, *MCL* mantle cell lymphoma, *MM* multiple myeloma

development of novel, unrelated drugs capable of inhibiting the proteasome (Dick and Fleming 2010). The peptide epoxyketone **Carfilzomib (Kyprolis)** is a derivative of the natural product epoxomicin isolated from Actinomyces. Carfilzomib inhibits the proteasome in a manner unrelated to Bortezomib. It irreversibly binds to the catalytic site and inhibits its activity. Carfilzomib is approved as a treatment for multiple myeloma since 2012. **Ixazomib (Ninlaro)** is the first approved oral proteasome inhibitor which reversibly inhibits the proteasome by a mechanism analogous to Bortezomib and has been approved in 2015 for the treatment of multiple myeloma in combination with lenalidomide and dexamethasone. Most proteasome inhibitors are of peptidic nature. The first non-peptidic proteasome inhibitor was the natural product **Lactacystin** which has been extensively used as a research tool, but has not been used to treat patients.

Targeted Monoclonal Antibodies
Targeted therapy using monoclonal antibodies (mAbs) takes advantage of the unique properties of antibodies to recognize specific cell surface proteins interrupting their downstream signaling pathways or inducing an immune response that destroys the cancer cells (Scott et al. 2012). Antibodies are glycoproteins of the immunoglobulin family of proteins used by our immune system to fight pathogens (Fig. 2.17).

They recognize the pathogens by specific molecules called antigens. The advent of hybridoma technology in the 1970s and more recently recombinant DNA technology, transgenic mice and phage display, allowed the production of mAbs by clonal cells providing a reliable source of antibodies that all bind to the same epitope. An epitope is a part of an antigen which is recognized by the antibody. Different types of mAbs have been developed, namely murine, chimeric, humanized and fully human. Most strategies of developing therapeutic antibodies involve generation in a non-human immune system mainly in mice. Initially murine antibodies produced by hybridoma technology were used for therapeutic purposes. In order to minimize their immunogenicity and to optimize their cytotoxic efficacy in humans their similarity to human antibodies has been increased by using different strategies. In order to generate **chimeric antibodies**, the antigen binding region from a murine antibody is fused with the constant, effector region from antibodies occurring naturally in humans resulting in an antibody that is approximately 65% human. **Humanized antibodies** can be engineered by recombinant DNA technology resulting in a protein that is approximately 95% human. **Fully human antibodies** can be produced by transferring human immunoglobulin genes into the genome of transgenic mice and

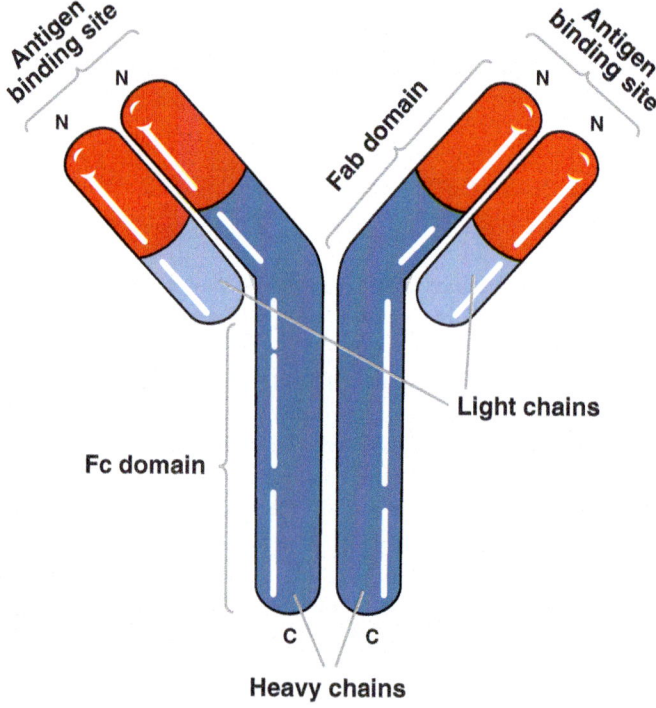

Fig. 2.17 A complete antibody consists of four polypeptides, two heavy chains and two light chains held together via disulfide bonds. Each polypeptide chain contains a constant region, that is very similar for all antibodies, and a variable region, that is specific to each particular antibody. The shape of an antibody is a Y. Antibodies are divided into several functional domains. The main portions are Fc (Fragment crystallisation) and Fab (Fragment antigen binding). Each antibody has two antigen-binding sites, and can bind to two antigens

challenging them with the desired antigen. Antibodies have become a fundamental part of targeted therapy as they interact with the target with a very high degree of specificity and hence avoiding unwanted target off effects (Box 8).

Box 8

On target and off target side effects
Side effects of a drug are defined as effects that are secondary to the intended therapeutic effect and mostly refer to adverse toxicologic effects. Side effects can be classified into on-target and off-target effects. A drug can cause toxicity by modulating its intended molecular target in normal tissue. This kind of adverse effects are called on-target or mechanism-based adverse effects. In this case the therapeutic target and the toxic target are identical. On-target side effects might lead to the identification of previously unknown functions

2.3 Chemical Treatment

of the therapeutic target. A typical on-target side effect of MEK inhibitors such as Trametinib is ocular toxicity. Immune checkpoint inhibitors can cause dermatologic toxicities such as rash, vitiligo or hair depigmentation in patients with melanoma by their intended activity, namely relieving the immunological brakes of immune cells to unleash an immune response against cancerous skin cells. Accordingly, these adverse effects have been reported to be a positive prognostic factor in these patients. Conversely, off-target adverse effects arise from interference with targets other than the intended therapeutic target. Very few drugs are thought to be absolutely specific for their intended molecular target. In order to investigate off-target adverse effects, genetically engineered mice in which the intended therapeutic target has been deleted can be treated with the drug. Any observed response to treatment must rely on targets different than those intended for the therapeutic effect. Tyrosine kinase inhibitors are known to be associated with an increased risk of cardiovascular events. As tyrosine kinases are involved in cardiomyocyte survival, their inhibition leads to on-target myocardial dysfunction. On the other hand, tyrosine kinases inhibit multiple kinases including not-intended kinases whose inhibition can result in off-target cardiac toxicities.

However, antibodies have also several disadvantages. They are large molecules that are unable to cross cellular membranes and therefore are limited to recognize their targets at the cell surface. Antibody-based drugs are generally delivered by intravenous or subcutaneous injection whereas oral administration of remains a major pharmaceutical challenge. The first mAb therapy was approved in 1986 for transplant rejection. The use of mAbs in clinical practice and in particular, against oncologic targets has become increasingly important ever since. We can distinguish **naked mAbs** that work by themselves either by binding to oncogenic surface receptors or by targeting immune checkpoints (discussed below in the section on Immunotherapy) or **conjugated mAbs** (discussed below in the section on Immunotoxins). Over 20 mAbs have been approved for the treatment of different cancers directed against tumor-associated surface proteins such as CD3, CD19, CD20, CD30, CD38 and CD52, oncogenic extracellular signaling molecules and receptors such as EGFR, VEGF, VEGFR, HER2 and PDGFR-α or immunocheckpoint molecules including CTLA-4, PD-1 and PD-L1.

mAbs Against Tumor-Associated Surface Proteins
Cancer cells express surface proteins that can be recognized by specific antibodies which in turn trigger their destruction by activating the complement system, inducing apoptosis or mediating antibody-dependent cellular cytotoxicity (Table 2.11). A CD (form "**Cluster of Differentiation**") number has been assigned to many of these surface proteins to identify the epitopes recognized by mAbs. Different cell population can be identified by their characteristic pattern of CD antigen expression. CD20 is a transmembrane protein of unknown function expressed on B-cells and can

Table 2.11 List of approved mAbs against tumor-associated surface proteins

Drug (Trade name)	Target	Indication
Rituximab (Rituxan)	CD20	CLL, NHL
Ofatumumab (Arzerra)	CD20	CLL
Obinutuzumab (Gazyva)	CD20	CLL, FL
Daratumumab (Darzalex)	CD38	MM
Dinutuximab (Unituxin)	GD2	Neuroblastoma
Blinatumomab (Blincyto)	CD19	ALL
Alemtuzumab (Campath)	CD52	CTCL, TCL
Elotuzumab (Empliciti)	SLAMF7	MM

ALL acute lymphoblastic leukemia, *CLL* chronic lymphocytic leukemia, *CTCL* cutaneous T cell lymphoma, *FL* follicular lymphoma, *MM* multiple myeloma, *NHL* non-Hodgkin's lymphoma, *TCL* T-cell lymphoma

be used as an antigen for therapeutic mAbs to destroy them. Therefore, therapeutic mAbs against CD20 have been used to get rid of B-cells under conditions where there are overactive or dysfunctional B-cells or an excessive number of them. **Rituximab (Rituxan)** is a chimeric mAb specific against CD20 used to treat several autoimmune diseases and types of cancers (Weiner 2010). **Rituximab** has been approved as early as 1997 and has been used for the treatment white blood cell cancers including leukemias and lymphomas. Rituximab eliminates normal and malignant B cells, but spares lymphoid stem cells which can repopulate healthy B cells. The clinical success of Rituximab prompted the development of second and third generations of anti-CD20 mAbs including the approved fully human mAbs **Ofatumumab (Arzerra)** and **Obinutuzumab (Gazyva)**. After the patent on Rituximab expired several **Biosimilars** (Box 9) have been approved recently (Mellstedt et al. 2008).

Box 9
Biosimilars
Biosimilar is the term used for an almost identical copy of a biopharmaceutical which can be produced when the patent on the original product expires. Biosimilars is analogous to the term "generic drug" which is used for a follow-on product of a small molecule drug. Different terms are used for copies of biopharmaceuticals and small molecule drugs as they involve different degrees of reproducibility. It is easier to copy a totally synthesized or semisynthetic chemical compound than recombinant therapeutic proteins such as monoclonal antibodies and hormones or gene- and cell-based therapies.

Daratumumab (Darzalex) is a mAb against CD38, another cell surface marker overexpressed by multiple myeloma cells and has been approved as the first mAb for the treatment of this disease in 2015. **Dinutuximab (Unituxin)**, targets the glycolipid GD2 expressed by cells of neuroectodermal origin including malignant neuroblastoma cells and is approved as a component of a combination regimen to treat neuroblastoma as a second line therapy. **Blinatumomab (Blincyto)** is an example of a bi-specific mAb that contains two distinct antigen binding sites. It represents a very elegant engineering effort that combines binding specificity against CD3 which is a part of the T cell receptor of T cells and CD19 expressed on B cells. The rationale behind this design is to link these two cell types and thereby activating T-cell mediated cell toxicity directed against B-cells. Accordingly, Blinatumomab is called a Bi-specific T-cell engager. Both antigens are expressed in children and adults and hence can be used for therapy in pediatric and adult patients. Blinatumomab has been approved for second line therapy of acute lymphoblastic leukemia (ALL) in 2014. **Alemtuzumab (Campath)** is a humanized mAb that recognizes CD52, a glycoprotein present on the surface of mature lymphocytes, monocytes and dentritic cells, but not expressed by the stem cells mature lymphocytes are derived from. Alemtuzumab destroys CD52-bearing cells by antibody-dependent cell-mediated cytotoxicity and has been approved for the treatment of B-cell chronic lymphocytic leukemia (B-CLL). **Elotuzumab (Empliciti)** is a humanized mAb directed against the surface antigen SLAMF7 (also called CD319) expressed on the surface of normal and malignant plasma cells and has been approved for the treatment of patients with multiple myeloma in 2015.

mAbs Against Oncogenic Extracellular Signaling Molecules and Receptors

Therapeutic mAbs that target extracellular signaling molecules and growth factor receptors on the surface of cancer cells have become a cornerstone of targeted therapy (Table 2.12). As in the case of mAbs that target CD surface markers of leukocytes (see above), mAbs binding to signaling proteins/receptors exert their anti-cancer activity by antibody-dependent cellular cytotoxicity. However, they are also capable

Table 2.12 List of approved mAbs against oncogenic extracellular signaling molecules and receptors

Drug (Trade name)	Target	Indication
Trastuzumab (Herceptin)	HER-2	Breast cancer
Pertuzumab (Perjeta)	HER-2	Breast cancer
Cetuximab (Erbitux)	EGFR	Colorectal cancer, NSCLC, HNC
Panitumumab (Vectibix)	EGFR	Colorectal cancer
Bevacizumab (Avastin)	VEGF-A	Colon and lung cancer, glioblastoma, RCC
Ramucirumab (Cyramza)	VEGFR2	GEJ, NSCLC
Olaratumab (Lartruvo)	PDGFRα	Soft-tissue sarcoma

GEJ Adenocarcinoma of the gastroesophageal junction, *HNC* head and neck cancer, *NSCLC* non-small cell lung cancer, *RCC* renal-cell carcinoma of interrupting the downstream signaling pathways activated by these proteins. The inhibiting effect of these mAbs on oncogenic cell signaling is considered as an important aspect of their efficacy. Most of these mAbs are directed against the extracellular domain of receptor tyrosine kinases such as HER2, EGFR, VEGFR or PDGFR. The paradigmatic example of such a drug is **Trastuzumab (Herceptin)**. Trastuzumab is a humanized mAb that binds to the HER2 receptor (Fig. 2.18; Valabrega et al. 2007).

HER-2 is a growth factor receptor which is required for cell growth in normal breast tissue. HER-2 is overexpressed in 30% of breast cancer patients either by transcriptional activation or gene amplification contributing to cancerous cell growth. Like other growth factor receptors HER2 dimerizes forming either homodimers with a second HER2 or heterodimers with other receptors including HER3. Trastuzumab binds to HER-2 at the cell surface at the domain required for HER2/HER2 dimerization and prevents HER-2 mediated growth stimulatory downstream signaling. In particular, Trastuzumab has been shown to downregulate the activity of the serine/threonine kinase AKT and stimulate the expression of the cell cycle inhibitory protein p27. As a result, disease progression is slowed down. However, 70% of breast cancer patients (with HER-2 negative tumors) would not benefit from the treatment with Trastuzumab which is expensive and associated with adverse effects. This is a

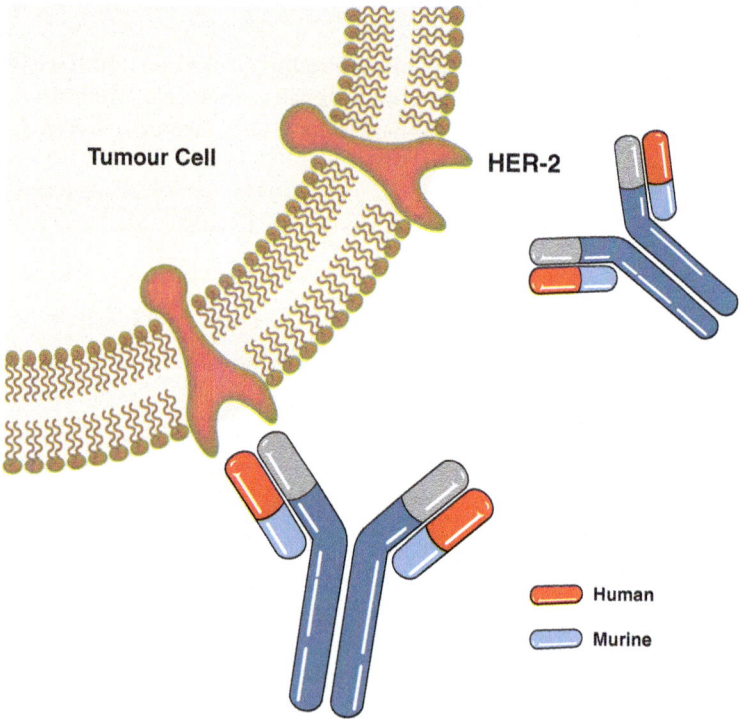

Fig. 2.18 Trastuzumab is a humanized monoclonal antibody that binds to the growth factor receptor HER2

2.3 Chemical Treatment

good example for the fact that many targeted therapies require companion diagnostic biomarkers to identify the subset of patients that would benefit from the corresponding targeted drug. In the case of Trastuzumab, several companion diagnostic tests that detect the overexpression of HER-2 by immunohistochemistry (IHC) and fluorescence in situ hybridization (FISH) have been approved by the US Food and Drug Administration (FDA). Trastuzumab is indicated for the treatment of patients with HER2 positive breast cancer and has been approved in 1998 (Jones and Buzdar 2009). Trastuzumab is an extremely successful drug whose patent has expired in Europe and will do so in the US in 2019, prompting many companies around the world to develop biosimilar versions of the drug. **Pertuzumab (Perjeta)** is humanized mAb that also binds to HER2 but at a different epitope. It recognized the domain necessary to form heterodimers with HER3. The HER2/HER3 heterodimer has been reported to be the most potent activator of downstream signaling. Accordingly, and Pertuzumab was coined a HER dimerization inhibitor. Pertuzumab has been approved for the treatment of metastatic HER2 positive tumors in combination with Trastuzumab and Docetaxel. The chimeric mAb **Cetuximab (Erbitux)** and the fully human mAb **Panitumumab (Vectibix)** recognize another growth factor receptor, namely EGFR. EGFR signalling has been shown to play an important role in the formation and progression of colorectal cancer. This type of cancer is highly prevalent and among the top three causes of cancer death in the Western world, both in men and women. Surgery and chemotherapy in a neo-adjuvant or adjuvant setting are the current cornerstones of CRC treatment. Furthermore, the management of metastatic CRC has significantly improved with the combination of chemotherapy with mAb-based targeted agents such as Cetuximab and Panitumumab. Cetuximab and Panitumumab bind to the extracellular domain of EGFR and interrupt downstream signalling. EGFR is expressed in the majority of patients with metastatic CRC. However, the expression of EGFR does not predict the response to anti-EGFR targeted therapy. Activating mutations in KRAS, a small GTP binding protein are present in over half of all colorectal cancers (CRC). As activated KRAS relays the signal to the downstream pathway (even if the upstream EGFR has been blocked by cetuximab), CRC tumors habouring these mutations correlate with a worse clinical response to Cetuximab therapy. Consequently, a PCR-based test to identify the presence of KRAS mutations was developed to select CRC patients with wild-type KRAS prior to Cetuximab treatment, representing the first genetic companion diagnostic test to guide the therapeutic strategy. Even with this companion screening, only a sub-group of CRC patients with wild-type KRAS benefit from anti-EGFR targeted therapy. As a result, a better understanding regarding the heterogeneity of CRC is required to allow a precise molecular stratification of CRC patients and improve the clinical prognosis of CRC patients. Bevacizumab and Ramucirumab are angiogenesis inhibitor which interfere with the VEGF/VEGFR signalling. **Bevacizumab (Avastin)** is a humanized mAb that binds to VEGF-A, a secreted, extracellular signaling molecule that binds as a dimer to its receptor VEGFR2 which dimerizes to propagate the signal promoting angiogenesis (see definition above). VEGFR2 has been characterized as the main stimulatory receptor of the VEGF molecules. By binding to this growth factor protein, Bevacizumab blocks VEGF signaling and thereby inhibits angiogenesis. Bevacizumab was the first angiogenesis inhibitor to be approved in 2004 for metastatic colon cancer in combination with standard chemotherapeutic drugs. In the following years Bevacizumab has been approved to treat several other solid tumors including lung cancer, glioblastoma, and renal-cell carcinoma always in combination with standard chemotherapy. The approval for breast cancer received in 2008 was revoked in 2011 due to lack of clinical benefit. The a fully human mAb **Ramucirumab (Cyramza)** isolated from a phage display library binds to the VEGFR2 thereby preventing the binding of the agonist VEGF and blocking signaling downstream of the receptor. Ramucirumab is indicated for second line treatment of gastric or gastro-esophageal junction adenocarcinoma and metastatic non-small-cell lung carcinoma in combination with standard chemotherapy. **Olaratumab (Lartruvo)** is a fully human mAb directed against the alpha subunit of the platelet-derived growth factor receptor (PDGFRα) thereby interrupting the propagation of the PDGF growth signals. Olaratumab gained regulatory approval for second line treatment of advanced soft-tissue sarcoma, a very aggressive cancer with limited treatment options.

Immunotoxins

mAbs can also be used as delivery vehicles, guiding radioactive molecules or toxins to the cancer cells to destroy them (Pastan et al. 2006). These antibody-drug conjugates are a type of immunotoxins. In general, immunotoxins consist of two distinct components, a targeting element and a toxin (Fig. 2.19).

The targeting moiety can be a modified antibody, an antibody fragment or a growth factor that specifically bind a surface protein expressed on the target cell. The targeting component can be fused or conjugated with a moiety that mediates cytotoxicity such as a cytotoxic protein derived from bacteria or plants, a small molecule poison or a cytotoxic radioisotope. The toxins can be connected to the targeting moiety by a

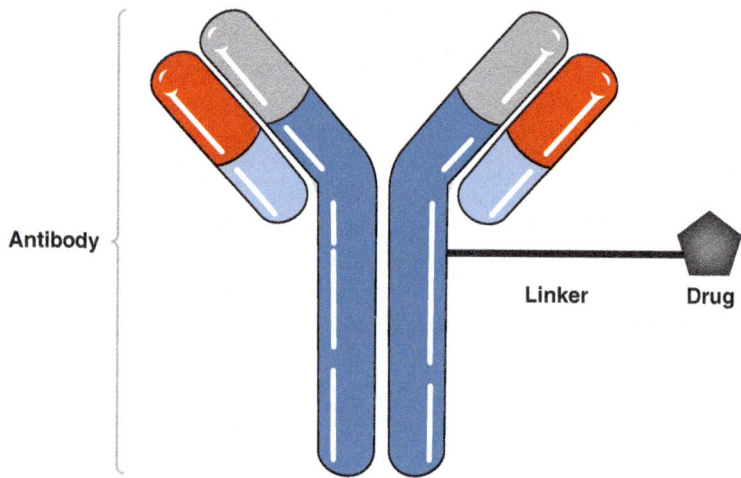

Fig. 2.19 Principles of Immunotoxins. Immunotoxins are artificial molecules consisting of a cytotoxic moiety linked (via a chemical linker) to a targeting portion

2.3 Chemical Treatment

linker molecule or produced as recombinant proteins. Upon binding of the targeting part to a specific cell surface protein, the toxin destroys the target cell by proximity to the target cell or after its uptake through endocytosis. Targets to guide the toxin to the cancer cell to be destroyed include to Interleukin-2 receptors, HER2, CD20, CD22, CD30 and CD33. The indication of immunotoxins depends on the cell population targeted by the targeting moiety. Approved immunotoxins are mainly used to treat hematological malignancies (Table 2.13). Solid tumors appear to be much more difficult to target by immunotoxins. Major challenges include ineffective penetration into the tissues of solid tumors and neutralizing activity of the immune system. The first approved immunotoxin was **Denileukin diftitox (Ontak)**, a chimeric protein consisting of interleukin-2 and the Diphtheria toxin. Interleukin-2 guides this immunotoxin to cells that bear Interleukin-2 receptors. Denileukin diftitox gained regulatory approval for the treatment of cutaneous T-cell lymphoma (CTCL) in 1999 but was withdrawn from the market in 2014. Similarly, **Gemtuzumab ozogamicin (Mylotarg)** approved as the first antibody-drug conjugate in 2000 was discontinued in 2010 due to its negative risk/benefit balance. Gemtuzumab ozogamicin combines a mAb specific for the CD33 surface marker and a cytotoxic agent called ozogamicin. Ozogamicin is a small chemical calicheamicin-type compound known to induce DNA damage. Ozogamicin is linked to the mAb by a small chemical linker molecule called AcBut. CD33 is present on normal hematopoietic cells and leukemic blast cells whereas hematopoietic stem cells do not express CD33. The antibody-drug conjugate binds to the CD33, enters the cell, where the toxin is cleaved from the mAb in the acidic environment of the lysosome. Once free, ozogamicin detroys the cell by damaging its DNA. Gemtuzumab ozogamicin was indicated for the treatment of relapsed acute myelogenous leukemia (AML). Recently it has been readmitted for the same indication. **Inotuzumab ozogamicin (Besponsa)** share with Gemtuzumab ozogamicin the toxin and the linker molecule but the targeting moiety is different. Inotuzumab ozogamicin links ozogamicin via AcBut to a mAb capable of directing the toxin to cells bearing the surface marker CD22. The targeting

Table 2.13 List of approved Immunotoxins

Drug (Trade name)	Target	Indication
Inotuzumab ozogamicin (Besponsa)	CD22	ALL
Rituximab and hyaluronidase human (Rituxan Hycela)	CD20	FL, DLBCL, CLL
Ibritumomab tiuxetan (Zevalin)	CD20	NHL
Brentuximab vedotin (Adcetris)	CD30	HL, ALCL
Trastuzumab emtansine (Kadcyla)	HER2	Breast cancer

ALCL anaplastic large cell lymphoma, *ALL* acute lymphoblastic leukemia, *CLL* chronic lymphocyticeukemia, *DLBCL* diffuse large B-cell lymphoma, *FL* follicular lymphoma, *HL* Hodgkin lymphoma, *NHL* non Hodgkin's lymphoma moiety of Inotuzumab ozogamicin is the humanized anti CD22 mAb inotuzumab. The antibody-drug conjugate binds to the CD22 cell surface receptor, which is highly expressed on B cells. The mechanism by which the target cell is destroyed is the same as in the case of Gemtuzumab ozogamicin. Inotuzumab ozogamicin is indicated for the treatment of CD22-positive B-cell precursor acute lymphoblastic leukemia (ALL). **Rituximab and hyaluronidase human** (Rituxan Hycela) combines the mAb Rituximab (see above) with the human endoglycosidase hyaluronidase. Hyaluronidase is an enzyme that reversibly depolymerizes hyaluronan, a polysaccharide of the extracellular matrix in subcutaneous tissues thereby increasing the tissue permeability locally. Rituximab guides the enzyme to normal and malignant B cells which express its antigen CD20. Accordingly, Rituximab and hyaluronidase human facilitates the penetration of co-administered drugs within the tumor tissue. Therefore, Rituximab and hyaluronidase human is not an immunotoxin in the strictest sense. The toxic component of this approach is a separate molecule which is administered in combination with the antibody-enzyme conjugate. Rituximab and hyaluronidase human obtained regulatory approval in 2017 for treatment of follicular lymphoma (FL), diffuse large B-cell lymphoma (DLBCL), and chronic lymphocytic leukemia (CLL) in adult patients in combination with standard chemotherapy. As Rituximab and hyaluronidase human, **Ibritumomab tiuxetan (Zevalin)** also uses the CD20 surface protein as a target antigen but the approach to destroy the target cell is completely different. Ibritumomab tiuxetan is an antibody-radioisotope conjugate consisting of the murine mAb Ibritumomab specific for CD20 and a radioactive isotope attached to the antibody by the chemical compound tiuxetan which is a chelator. The isotope added to Ibritumomab tiuxetan is yttrium-90 or indium-111. As in the case of Rituximab, Ofatumumab or Obinutuzumab, Ibritumomab recognizes the CD20 antigen on the surface of normal and malignant B cells. The attached radioactive isotope destroys the cells that are close enough to the damaging radiation. In addition, the mAb component of the conjugate might act against B-cells itself by activating the complement system or antibody-dependent cell-mediated cytotoxicity. Ibritumomab tiuxetan was the **first radio immunotherapeutic drug** that obtained regulatory approval in 2002. It is indicated for the treatment of non-Hodgkin's lymphoma. **Brentuximab vedotin (Adcetris)** is an antibody-drug conjugate with a sophisticated design. It combines the chimeric mAb Brentuximab directed against the CD30 surface antigen with four different chemical groups: an attachment group, a cleavable linker, a space molecule and a toxin. Three to five molecules of the cytotoxic small molecule agent monomethyl auristatin E (MMAE) are linked by a small chemical spacer molecule to a peptide linker consisting of citrulline and valine. Brentuximab guides the conjugate to CD30 bearing cells which internalize Brentuximab vedotin. Once inside the target cell the peptide linker is cleaved by proteases in the lysosome and the toxin is released and can start to destroy the cell. As CD30 surface marker is a hallmark of Hodgkin lymphoma (HL) and systemic anaplastic large cell lymphoma (ALCL), Brentuximab vedotin obtained regulatory approval for these indications. The only antibody-drug conjugate approved to treat a solid cancer is **Trastuzumab emtansine (Kadcyla)** (Verma et al. 2012). It combines the mAb Trastuzumab that recognizes HER2 (as discussed above) with the small molecule

2.3 Chemical Treatment

spindle poison emtansine. The mAb is linked to several molecules of emtansine by a small chemical linker. Trastuzumab emtansine is thought to kill HER2 bearing target cell in two ways. After binding to the HER2 receptor the antibody-drug conjugate is internalized by endocytosis and emtansine released to interfere with cell division (see the section on spindle poisons). However, the mAb component of the conjugate, Trastuzumab can act on its own to inhibit HER2 downstream signalling and trigger antibody-mediated cell toxicity as discussed above. Trastuzumab emtansine has been approved to treat HER2-positive metastatic breast cancer in 2013.

The Limitations of Targeted Therapies
Targeted therapies have been introduced in recent years and at present the impact is limited to some specific types of cancer. These are still early days to judge whether targeted therapies will mark a true breakthrough in cancer treatment. The widespread optimism is not shared by everyone, however (Kim 2003). It has been argued that most targeted therapies offer only marginal extensions of life and few cures. Considering the enormous costs of these treatments, gains are rather modest. Some researchers suggest that we should focus more on metabolic and oxidative vulnerabilities that arise as a consequence of the uncontrolled growth and proliferation capacities of all cancer cells, rather then on targeting molecular events specific only for a small subset of a given cancer type. It is important to note that intrinsic or acquired resistance still limits the efficacy of targeted therapies in cancer treatment. Selective pressure in combination with mutations, epigenetic alterations or changes in microenvironment lead to resistant cancer cells and in turn to tumor regrowth and clinical relapse. As the malignant phenotype is often regulated by multiple parallel pathways the cancer cell may start to use alternative rescue signaling, if the main route has been targeted by an inhibitor. Therefore, it might be useful to block several supporting pathways using combination therapies with other anticancer agents to prevent resistance development. Importantly, the determination of resistance mechanisms can provide the basis for the design of second-generation therapies. This strategy has been successfully employed to inhibit BCR-ABL with Imatinib resistant point mutations using the second-generation kinase inhibitor Dasatinib.

2.4 Biological Treatment: E.g. Immunotherapy and Oncolytic Viruses

Biological treatment options for human cancers refer to therapies involving the use of antibodies, recombinant proteins, viruses and cells. Here we discuss the use of mAb and cells for immunotherapy and oncolytic viruses. mAbs used to interfere with signalling pathways or to guide toxins towards target cells have been discussed in the context of targeted therapies though their anti-cancer activity is partially due to activating the host immune system.

2.4.1 Immunotherapy

Immunotherapy is the treatment of diseases by stimulating or suppressing an immune response. Cancer immunotherapy or immuno-oncology refers to a treatment that stimulates the immune system of the patient enabling a strong immunological response against the tumor. Our immune system can be classified into **innate and adaptive immune system** (Fig. 2.20) and consists of cells, tissues and molecules that protects us from microbial pathogens, but is not optimized to respond to cancer, as cancer cells arise from our own cells.

Nevertheless, even cancer cells can be attacked by the immune system because they express so called **tumor antigens** (Box 10) (Schumacher and Schreiber 2015).

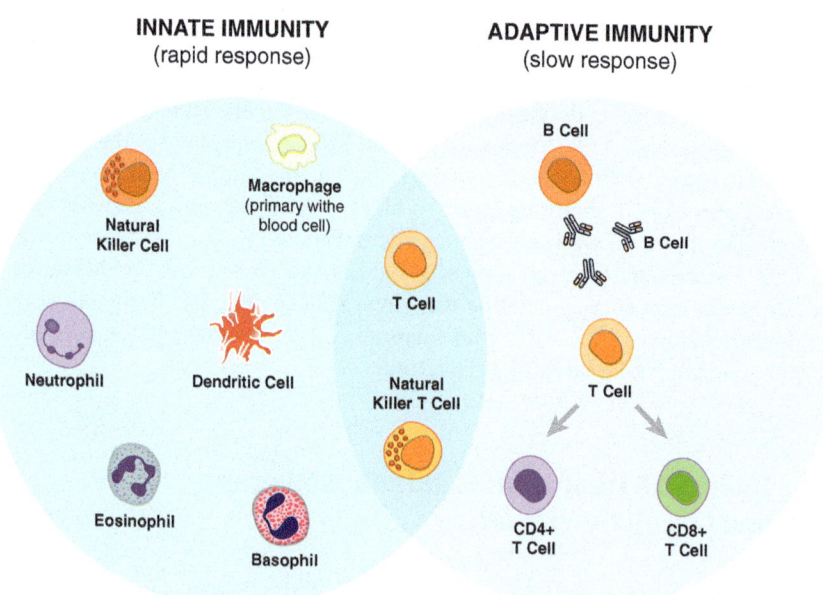

Fig. 2.20 Innate and adaptive immune system. The innate immune system constitutes a rapid first line of defense based on soluble factors including complement proteins, and diverse cellular components including granulocytes (basophils, eosinophils and neutrophils), mast cells, macrophages, dendritic cells and natural killer cells. The adaptive immune system is more specific but slower and is based on antibodies, B cells, and CD4+ and CD8+ T lymphocytes. Innate and adaptive immune system work closely together via cytotoxic lymphocytes, namely natural killer T cells and $\gamma\delta$ T cells. Adapted from Nature Reviews Cancer 4, 11

> **Box 10**
> **Tumor antigens**
> A tumor antigen is a substance produced by tumor cells that can be recognized by the immune system and initiate an immune response in the host. Normal cells usually don't generate antigenic substances because they are recognized as self-produced antigens. This important property of our immune system is called self-tolerance, a process by which self-reactive lymphocytes namely cytotoxic T cells and autoantibody-producing B cells are removed from the repertoire before they can trigger damaging autoimmune reactions. However, our immune system might recognize proteins produced by our own cells to which it has not been exposed before. This can happen if a normal protein is expressed at extremely low level, or only at a certain stage of development or sequestered away from the immune system. Another possibility is that a protein suffers a modification in its structure due to a mutation. Tumor antigens have been classified into **tumor-specific antigens** (TSAs) and **tumor-associated antigens** (TAAs). TSAs are only present in tumor cells whereas TAAs are predominantly present in tumor cells, but also in normal cells. These antigens can be cell surface proteins or small parts of intracellular molecules presented on the cell surface as part of the major histocompatibility complex. An important group of tumor antigens are the **oncofetal antigens**. These proteins are usually expressed in early stages of embryonic development when the immune system is still in mature but disappear before self-tolerance develops. Another important class of tumor antigens are the neo-antigens. **Neo-antigens** arise from genes with somatic mutations in tumor tissue leading to altered amino acid coding sequences. It has been shown that mutations in the driver genes BRAF, RAS, and PIK3CA can lead to the generation of neo-antigens. Importantly, passenger missense mutations can also lead to the generation of tumor-specific neo-antigens. In fact, the great majority of tumor antigens represent molecules that do not participate causally in neoplastic transformation. Many neo-antigens have been identified in melanoma, due to the relatively high mutation load in this tumor type. In addition, genetic and epigenetic alterations in a tumor can result in the abnormal high levels of proteins which are usually absent or expressed in very low quantities including melanoma-associated antigens or prostate-specific antigens.

Scientists did not give up the hope to use the immune system to treat cancer for the last hundred years and only recently immunotherapy has turned out to be one of the most promising approaches to fight human cancer. The concept of immunotherapy was based on the early observation that concomitant bacterial infection caused spontaneous remission in patients with usually incurable tumors. At the end of the nineteenth century the American surgeon William B. Coley successfully treated cancer patients injecting heat-killed bacteria. But it took a long time to understand the underlying mechanism of this clinical result and it was not before the 1960s when modern immunology started to generate the groundwork for today's possibilities to modulate immune functions. Important concepts in immune-oncology are immune-surveillance and cancer immunoediting (Box 11) (Swann and Smyth 2007).

> **Box 11**
> **Immune-surveillance and cancer immunoediting**
> Immune-surveillance is the ability of the immune system to identify and eliminate tumor cells. The immune-surveillance concept implies that cancerous cells arise frequently but are recognized and destroyed by the immune system before they can harm the host in a similar manner than invading pathogens. Immune-surveillance is thought to efficiently act against emerging cancer cells, because they express tumor antigens (see Box 10), but it is evident that our immune system is not very good at eradicating established tumors. The concept of cancer immunoediting tries to explain this observation by defining the relationship between tumor cells and the immune system. The idea is that the anti-tumor response of the immune system eventually leads to a change in the immunogenicity of the tumor and to immune resistant tumors. Immunoediting is characterized by three phases elimination, equilibrium, and escape. The elimination phase refers to the process described above as immune-surveillance. During this first phase, immune effector cells including macrophages, natural killer cells and cytotoxic T cells eliminate cells that bear tumor antigens. Elimination of these cells can lead to the complete clearance or might destroy only a portion of tumor cells. If the elimination of tumor cells is incomplete, a temporary equilibrium between the the developing tumor and the immune system can evolve. During this second phase of immunoediting which is called the equilibrium phase, immune selection mediated by lymphocytes and interferone-gamma keeps tumor cells dormant or reduces immunogenicity of tumor variants by acquiring further genetic or epigenetic changes. During the equilibrium phase the pressure exerted by the immune system prevents control tumor progression. However, tumor cell variants that manage to modulate their tumor antigens might become resistant to immune effector cells and enter the third phase called escape phase. In this last phase of immunoediting, resistant tumor cells escape from the immune attack and grow in an uncontrolled fashion leading to a growing tumor.

Immune-based treatments for cancer encompass tumor-specific mAbs, therapeutic cancer vaccines, cell therapies, cytokines and immune checkpoint inhibitors. These treatments can be classified into passive and active immunotherapy (Fig. 2.21). Whereas **passive immunotherapy** delivers already active immune factors usually mAbs, cytokines or antigen specific adaptive immune cells, **active immunotherapy** is aimed at activates intrinsic factors of the patient's own immune system against the tumor through vaccination or immune checkpoint inhibitors. Within this section

2.4 Biological Treatment: E.g. Immunotherapy and Oncolytic Viruses

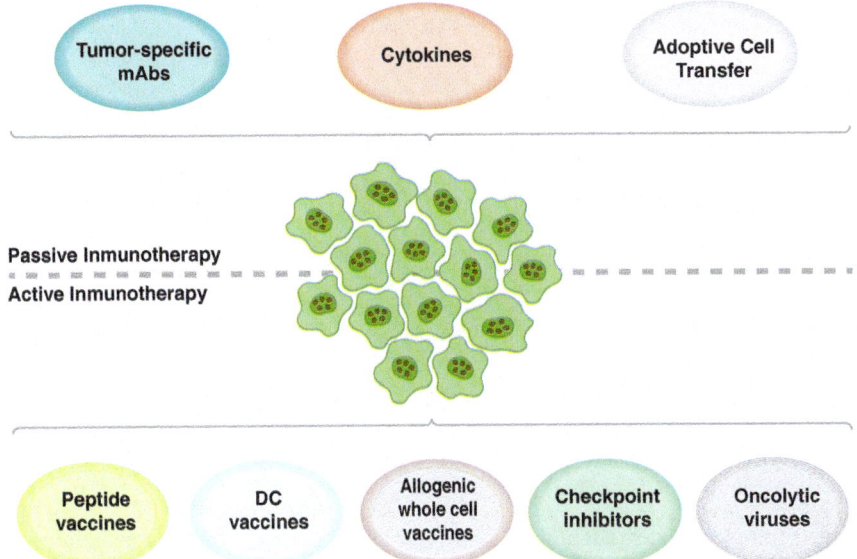

Fig. 2.21 Immunotherapy can be classified into passive and active immunotherapy. Passive immunotherapy delivers already active immune factors such as antibodies, cytokines or immune cells whereas active immunotherapy employs vaccines, viruses or immune checkpoint inhibitors to activate the immune system of the host to attack the tumor. Adapted from Ann Transl Med 4, 261

we will discuss cytokine therapy, cellular immunotherapy and immune checkpoint therapy and dedicate a separate section to oncolytic viruses.

Cytokine Therapy

Cytokines are a group of small extracellular signalling proteins which act through binding to cytokine receptors activating the JAK-STAT signalling pathway and are known to regulate inflammatory and immune responses (Fig. 2.22). Cytokines can be used as passive cancer treatments. They elicit a non-specific immune response not directed against a specific antigen. Immuno-stimulatory cytokines such as Interleukin-2 and Interferon-α stimulate a broad-based immune response.

Interleukin-2 (Proleukin) binds to receptors on the surface of T cells stimulating them to proliferate and to produce cytokines which mediate the activation of multiple types of immune cells including effector T cells and T-regulatory cells. High-dose of recombinant human Interleukin-2 has been approved to treat patients with renal cell carcinoma and melanoma in 1992 and 1998, respectively and can lead to a complete response in a small percentage of these patients. Accordingly, Interleukin-2 therapy is able to successfully manipulate the endogenous anti-tumor immune response in some patients. **Interferon-α (Multiferon)** is a pleiotropic cytokine which promotes the differentiation and activity of host immune cells. Together with Interferon-β, Interferon-α belongs to the type I Interferons which are produced as a general response of mammalian cells to a viral infection and help block viral replication in multiple ways.

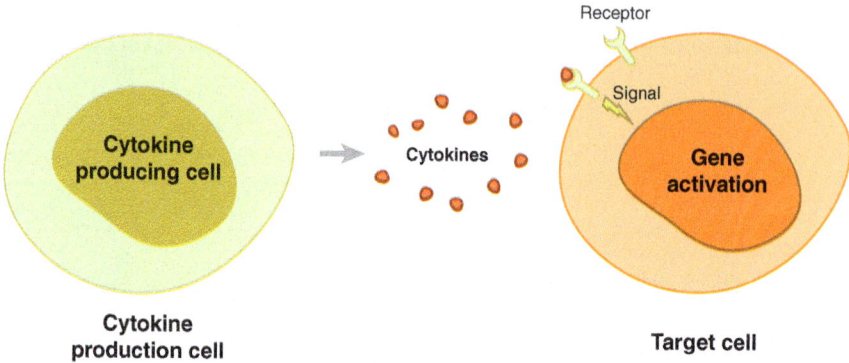

Fig. 2.22 Mode of action of cytokines. Cytokines are small extracellular signalling proteins which bind to cytokine receptors activating intracellular signalling and eventually regulating gene expression

They promote the degradation of RNA and block protein synthesis in uninfected neighboring cells, induce apoptosis in virus-infected cells and enhance the activity of natural killer cells. The molecular mechanism of its anti-tumor activity remains to be elucidated but is thought to be mediated by a modulatory effect on immune cell populations or a direct effect on viability, proliferation or composition of tumor cells. Interferon-α has been used to treat patients with hairy-cell leukaemia, AIDS-related kaposi's sarcoma, follicular lymphoma and chronic myeloid leukaemia.

Cellular Immunotherapy

Adoptive cell therapy is a passive immunotherapy modality based on the in vitro expansion, activation or manipulation of cells from the patient which are subsequently injected back into the patient to attack the tumor (Rosenberg et al. 2008). Many different strategies for adoptive cell therapy have been explored including removal of immune cells from the patient's blood or directly from the tumor, using different types of immune cells such as cytotoxic T cells, dendritic cells or natural killer cells or using genetically engineered immune cells. For the time being, two approaches, dendritic cell therapy and CAR T-cell therapy lead to approved treatment options used in the clinical practice to treat cancer (Table 2.14).

Table 2.14 List of approved adoptive cell therapies

Drug (Trade name)	Target	Indication
Sipuleucel-T (Provenge)	PAP	Prostate cancer
Axicabtagene ciloleucel (Yescarta)	CD19	LBCL
Tisagenlecleucel (Kymriah)	CD19	ALL

ALL acute lymphoblastic leukemia, *LBCL* diffuse large B-cell lymphoma, *PAP* prostatic acid phosphatase

2.4 Biological Treatment: E.g. Immunotherapy and Oncolytic Viruses

Dendritic Cell Therapy

Dendritic cell therapy is a cellular immunotherapy which induces dendritic cells to present tumor antigens to lymphocytes which in turn are activated to destroy cancer cells which present the same tumor antigen on their surface (Palucka and Banchereau 2012). Dendritic cells act as messengers between the innate and the adaptive immune systems. Dendritic cells express **Pattern Recognition Receptors** (Box 12) which enable them to recognize and phagocytose invading pathogens.

Box 12

Pattern Recognition Receptors

Non-epithelial cells including macrophages sense the presence of pathogens through Pattern Recognition Receptors (PRRs). PRRs are transmembrane or intracellular receptors which act as an alarm system to alert the innate and adaptive immune systems of an infection. PRR class of proteins encompasses a broad variety of receptor proteins belonging to the Toll-like receptors, the Nod-like receptors or the C-type lectin receptors. PRRs bind nucleic acids, lipids, polysaccharide and proteins which are typically present in pathogens called pathogen-associated molecular patterns (PAMPs). PAMPs are molecules specific for pathogens not present in normal cells in the body such as double-stranded RNA, bacterial liposaccharide (LPS), Flagellin, unmethylated CpG DNA, β-Glucan or degradation products of peptidoglycans. When PRR binds a PAMP, it stimulates the cell to secrete extracellular signal molecules including Prostaglandins and Cytokines to trigger an inflammatory response at the site of infection. Inflammation promotes the movement of cells from the blood to the infected tissue. Some of these cells are phagocytic cells that engulf and destroy pathogens.

Activated dendritic cells cleave the proteins of the phagocytosed pathogen into peptide fragments which bind to newly synthesized major histocompatibility complex (MHC) proteins which then carry the fragments to the dendritic cell surface. Dendritic cells present these fragments or antigens to the adaptive immune systems. This is, why dendritic cells are called antigen presenting cells. Antigen presenting cells activate lymphocytes preparing them to kill other cells that present the antigen. Likewise, dendritic cells can be induced to present tumor antigens priming lymphocytes to destroy cancer cells that express the tumor antigen. Different strategies to activate dendritic cells can be used including therapeutic vaccination (in contrast to prophylactic vaccination) in vivo or ex vivo activation. In order to activate dendritic cells by **vaccination in vivo**, lysates from the patient's own tumor or small peptides of the tumor antigen are usually given together with highly immunogenic substances so called adjuvants. These adjuvants usually consist of killed microbial material and trick the immune system making it think that the tumor antigens are bacterial or viral antigens. Once injected, dendritic cells process the antigen and express antigenic fragments on their surface. The activated dendritic cells then promote a T-cell based immune response against cancer cells that carry the tumor antigen. A second approach to activate dendritic cells is to remove them from the blood of a patient and activate them outside the body and reinfuse the activated cells back into the patient. **Ex vivo activation** of dendritic cells is carried out in cell cultures in the presence of a specific tumor antigen or a tumor lysate. The activated dendritic cells are usually injected back into the patient together with strong adjuvants. **Sipuleucel-T (Provenge)** is the only dendritic cell therapy that received regulatory approval for clinical practise (Kantoff et al. 2010). Sipuleucel-T is a therapeutic vaccination that involves the ex vivo activation of dendritic cells previously isolated from the blood of a patient. The cultured dendritic cells are exposed to antigen prostatic acid phosphatase (PAP), a protein which is present in almost all prostate cancer cells and to granulocyte-macrophage colony stimulating factor (GM-CSF) a which is an immune-stimulatory cytokine known to activate dendritic cells. The activated dendritic cells are then reinfused into the donor patient to attack and kill prostate cancer cells that express PAP. This procedure is repeated three times and has been approved in 2010 for the treatment of metastatic, hormone-refractory prostate cancer.

CAR T-Cell Therapy

CAR T-cell therapy is a cellular immunotherapy in which T cells are genetically engineered ex vivo to express a chimeric antigen receptors (CAR) specific for an arbitrary antigen (Kershaw et al. 2013). In this therapeutic modality, T-cells from a patient are changed in such a way that they will attack cancer cells (Fig. 2.23). T cells can only recognize fragments of protein antigens that have been produced by partial proteolysis inside a host cell. T cells bind antigens displayed on the host cell by **Major Histocompatibility Complex** (MHC) proteins via their T-cell receptors and mediate an antigen-specific immune response (Trowsdale and Knight 2013). T cells are only activated if they bind with high affinity to antigens/MHC and receive co-stimulatory signals. Activated cytotoxic T cell can recognize any target cell harbouring the same antigen and kill them by Fas ligand/Fas receptor pathway or the perforin/granzyme system. T-cell mediated cytotoxicity can target intracellular antigens (because they are processed and presented by MHC proteins). CAR T-cell system shortcut this complex process of T-cell activation by manipulating the T cell receptor. T cells isolated from a cancer patient are genetically engineered to express an artificial T cell receptor containing parts of different proteins. This so called chimeric antigen receptor is composed of a single chain antibody (scFv) directed against a specific tumor antigen fused to intracellular signalling adaptors for T cell receptor signalling. Second and third generations of CARs also contain sequences from costimulatory proteins which promote the expansion of T cells.

Axicabtagene ciloleucel (Yescarta) is a CAR T-cell therapy in which the T cells isolated from a patient is engineered to express an artificial T cell receptor directed against CD19 receptors on B cells (Neelapu et al. 2017). Axicabtagene ciloleucel has obtained regulatory approval to treat certain cases of large B cell lymphoma in 2017. At the same time a second CAR T-cell therapy targeting the same antigen has been approved for a related indication. **Tisagenlecleucel (Kymriah)** targets CD19 and is indicated for the treatment of B-cell acute lymphoblastic leukemia (ALL) (Maude

2.4 Biological Treatment: E.g. Immunotherapy and Oncolytic Viruses

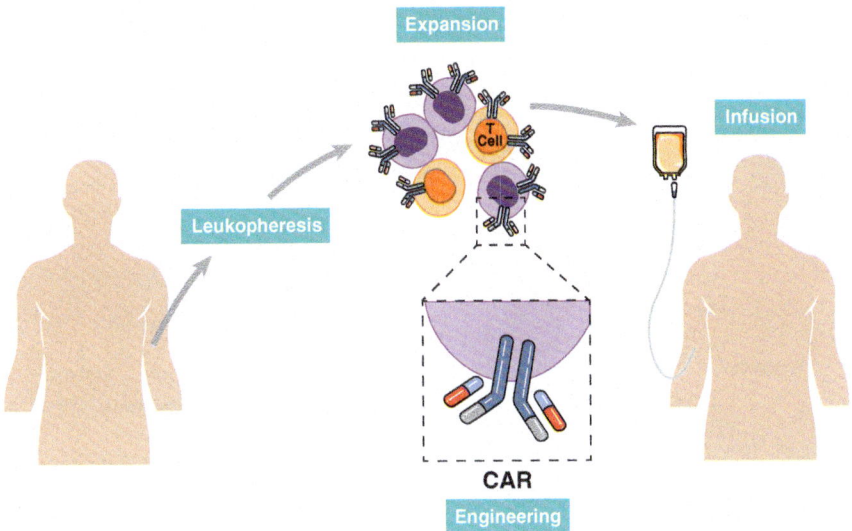

Fig. 2.23 CAR-T cell therapy uses engineered T cells with artificial T cell receptors aimed at targeting a specific tumor antigen. White blood cells from the blood of a patient are isolated (Leukophoresis) and genetically engineered in the laboratory in order to express a chimeric antigen receptor (CAR) in T cells. The resulting CAR T cells are expanded and then given to the patient by infusion. The CAR T cells can bind to a specific antigen on the cancer cells and kill them

et al. 2018). The identification of specific tumor antigens which are not present in normal cells is one of the biggest challenges associated with this kind of therapeutic approaches.

Immune Checkpoint Therapy
Immune checkpoint therapy consists in blocking inhibitory costimulatory molecules to unleash anti-tumor T cell responses (Sharma and Allison 2015). Immune checkpoint inhibitors relieve the immunological brakes which have evolved to avoid overactivation of the immune system on healthy cells. Immune checkpoints are very important to maintain the balance between stimulatory and inhibitory inputs to regulate an efficient response but at the same time preventing an overshooting which might lead to auto-immune responses. As noted earlier, T cells become induced by cytokines released from activated T-cells as part activated by binding of its T cell receptor to an antigen presented by a MHC protein on an antigen presenting cell. The full activation of the T cell receptor requires the interaction of the costimulatory molecules CD28 present on T cells and its ligands B7-1 and B7-2. However, the activation is also regulated by inhibitory checkpoints molecules known to be of a negative feedback loop to control excessive tissue damage. Tumor cells often take advantage of these checkpoints to protect themselves from a T cell attack by expressing immune checkpoint molecules. CTLA-4 and PD-1 are immune checkpoint proteins that have been studied as targets for cancer therapy (Fig. 2.24). **T-lymphocyte-associated protein 4 (CTLA-4)** is a CD28 homolog present on activated T cells and has been

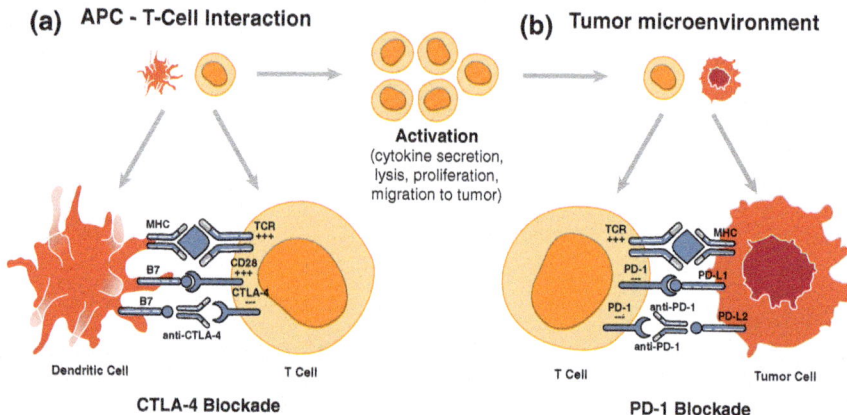

Fig. 2.24 Immune checkpoint therapy. Immune checkpoint inhibitors target regulators of the immune system that inhibit immune responses. Tumors can use these checkpoints to escape from a escape from an attack of the immune system. Currently, inhibitors are approved to target the molecules CTLA4 (**a**), PD-1 (**b**), and PD-L1. CTLA4 binds to B7-1 and B7-2 whereas PD-1 interacts with PD-L1

discovered in the 1980s. CTLA-4 was shown to compete with CD28 for B7 binding, but with higher affinity and negative downstream signalling (Fig. 2.24a) limiting their stimulatory effect of CD28 on T cells (Postow et al. 2015). Therefore, the focus of cancer immunotherapy shifted from activation T cell response to releasing its brakes by blocking the inhibitory signal of CTLA-4. In the 1990s, mAbs capable of blocking CTLA-4 were successfully used for the treatment of tumors in immunocompetent mice. The accumulated knowledge about the complexity of the T cell activation and its negative feedback loop was translated to the clinic in the 2000s. **Ipilimumab (Yervoy)** is a fully human mAb against CTLA-4 and was the first immune checkpoint inhibitor to be approved in 2011 for its therapeutic use in humans (Hodi et al. 2010). Ipilimumab mediated blockage of CTLA-4 improves the survival of about 20% of patients with advanced melanoma, a very aggressive cancer resistant to all standard treatment options. The clinical success of Ipilimumab promoted the emerging field of immune-checkpoint therapy. Another checkpoint molecule was cloned in 1992 and called **programmed death-1 (PD-1)** as it was shown to play a role in negative selection of T cells by programmed cell death (Postow et al. 2015). As CTLA-4, PD-1 is expressed in activated T cells but attenuates T cell activity by a distinct molecular mechanism than CTLA-4. PD-1 does not interfere with co-stimulation but rather inhibits downstream signalling by dephosphorylation of T cell receptor proximal signaling components (Fig. 2.24b).

Two ligands of PD-1 have been identified, PD-L1 and PD-L2. Whereas PD-L1 is expressed in many cell types, PD-L2 expression is more restricted to antigen presenting cells. PD-L1 protects tumor cells by inducing T cell apoptosis and its expression can be induced in tumors to evade an immune attack. Therefore, the PD-1/PD-L1 pathway is considered as an adaptive immune resistance mechanism

2.4 Biological Treatment: E.g. Immunotherapy and Oncolytic Viruses

Table 2.15 List of approved immune checkpoint inhibitors

Drug (Trade name)	Target	Indication
Ipilimumab (Yervoy)	CTLA-4	Melaoma, RCC
Nivolumab (Opdivo)	PD-1	Melaoma, NSCLC, RCC, Bladder cancer.
Pembrolizumab (Keytruda)	PD-1	Melaoma, NSCLC, HL, HNSCC
Atezolizumab (Tecentriq)	PD-L1	Melaoma, NSCLC, Bladder cancer.
Avelumab (Bavencio)	PD-L1	MCC, Bladder cancer
Durvalumab (Imfinzi)	PD-L1	Bladder cancer

HNSCC head and neck squamous cell carcinoma, *MCC* Merkel-cell carcinoma, *NSCLC* non-small cell lung cancer, *RCC* renal cell carcinoma

triggered by tumor cells upon an endogenous anti-tumor immune response. As in the case of CTLA-4, blockade of the inhibitory activity of PD-1 and PD-L1 was translated to the clinic by the development of specific antibodies that were shown unleash activated tumor-reactive T cells (Table 2.15). In clinical trials several of these antibodies have induced durable anti-tumor immune responses in a subset of patients in an increasing number of cancer types. These successful clinical results resulted in the approval of several antibodies against PD-1 and PD-L1 for their clinical use to treat several types of cancer. **Nivolumab (Opdivo)** was the first mAb against PD-1 approved for its clinical use. Nivolumab is a fully human mAb which was initially approved for the treatment of metastatic melanoma in 2014 (Robert et al. 2015b). Subsequently, its efficacy against squamous non-small cell lung cancer, renal cell carcinoma, colorectal cancer, hepatocellular carcinoma and classical Hodgkin lymphoma was shown and approved for these indications. **Pembrolizumab (Keytruda)** is a humanized mAb also directed against PD-1 and was approved for the treatment of metastatic melanoma in 2014 and a year later to treat metastatic non-small cell lung cancer (NSCLC) patients. For the time being three antibodies against PD-L1 capable of blocking the interaction between PD-1 and PD-L1 and in turn abrogating the inhibitory effect on T cell receptor signalling have been approved for their use in clinical practice. The first one was the humanized mAb **Atezolizumab (Tecentriq)** indicated for the treatment of bladder cancer and approved in 2016. The fully human mAb **Avelumab (Bavencio)** is approved to treat metastatic Merkel cell carcinoma. **Durvalumab (Imfinzi)** is also a fully human mAb approved for the treatment of bladder and non-small cell lung cancer. The development of antibodies against PD1- and PD-L1 is most actively pursued in the field of immune-checkpoint inhibitors as they have shown to cause less adverse effects than anti-CTLA-4 treatments. Currently more than 40 different antibodies against PD-1 and PD-L1 are being tested in clinical trial for the treatment of many different human malignancies. Despite unprecedented clinical responses, immune-checkpoint inhibitors face several major challenges. Usually less than 30% of patients respond to immune checkpoint inhibition therapy, meaning that more than 70% of patients do not benefit, and it is difficult to accurately predict the response. mAbs against PD-1 and CTLA-4 are more effective in tumors with high mutation rates which are more immunogenic and in tumors that are infiltrated by T cells. Some tumors with high expression level of PD-L1 have been shown to respond better to anti-PD-L1 treatment. As the mechanisms underlying the inhibitory effect of CTLA-4 and PD-1 is different, their inhibition by the corresponding mAbs interfere with different pathways. Therefore, anti-CTLA-4 and anti-PD-1 antibodies are being tested in combination therapy.

2.4.2 Oncolytic Viruses

Oncolytic viruses are viruses that preferentially infect and destroy cancer cells (Fig. 2.25; Chiocca 2002). The idea to treat cancer with viruses is based on the early observation of spontaneous cancer regression after viral infections. The preference of oncolytic viruses towards cancer cells is based on the selective growth advantages for viruses in cancer cells compared to normal cells. Cancer cells proliferate faster and are preferable targets for viral replication. Furthermore, cancer cells frequently loose critical anti-viral defence mechanisms such as the interferon response system is which is disrupted in many transformed cells.

The obvious problem associated with virotherapy is the patient's own immune system which might eliminate the virus before it can attack the tumor. Early human trials using viruses often resulted in tumor regression but was followed by tumor progression during the late stages of these trials and sometimes cases of dangerous uncontrolled infection occurred. Therefore, the scientific community lost interest in the field of cancer virotherapy and it was not until the advent of recombinant DNA technology providing the required methods to modify viruses at will, that this ther-

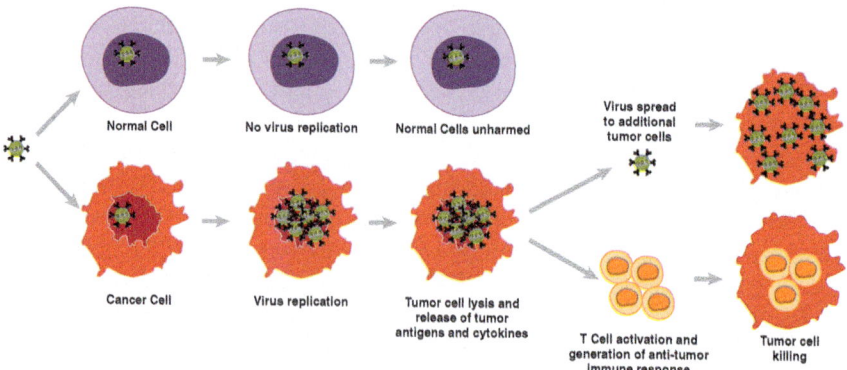

Fig. 2.25 Oncolytic viruses preferentially infect and destroy cancer cells leaving normal cells unharmed. Adapted from EMBO Mol Med e201607296

2.4 Biological Treatment: E.g. Immunotherapy and Oncolytic Viruses

apeutic option attracted attention again. Genetic engineering of oncolytic viruses is aimed at eliminating non-essential viral genes increasing safety, oncolytic activity and specificity for cancer cells. **H101 (Oncorine)** is a genetically modified adenovirus which carries a loss-of-function mutation in E1B, a protein that inactivates p53. Therefore, H101 will readily replicate and cause cell death in cells with mutated p53, but not in p53 wild type cells. H101 was approved in 2005 by China's State Food and Drug Administration (SFDA) for the treatment of head and neck cancer. **Talimogene laherparepvec or T-vec (Imlygic)** is a genetically engineered, attenuated herpes simplex virus 1 (HSV1) from which two viral genes were removed and one human gene added (Andtbacka et al. 2015). T-vec is unable to replicate in normal cells due to the removal of the viral gene that codes for infected cell protein 34.5 (ICP34.5). ICP34.5 abrogates a cellular anti-viral defense mechanism, known as stress response that leads to cell death of infected cells. As outlined above, in many cancer cells anti-viral defense mechanisms are already interrupted by oncogenic activity and therefore the virus does not rely on ICP34.5 for its replication. Therefore T-vec preferentially replicates in cancer cells. After successful replication and the production of virus particles the host cell finally bursts releasing intracellular tumor antigens and virus which can infect and kill other tumor cells. T-vec also carries the human gene coding for the stimulatory cytokine GM-CSF that activates dendritic cells. Dendritic cells can swallow tumor antigens released from the tumor and present them on their surface to activate a T cell mediate immune response. The second viral gene removed from the HSV1 was the gene coding for ICP47, a protein that helps the virus to evade immune destruction by suppressing the host immune response. The T-vec design is aimed at a double hit consisting in direct destruction of the tumor cell by the virus and the induction of an immune response against the tumor antigens. T-vec is administered locally by injecting it into the tumor. T-vec has been approved in 2015 as the first onlolytic virus that obtained permission to be used in clinical practice in the western world. T-vec is used to treat melanoma lesions that cannot be operated.

Thought Questions

1. How can you classify the modalities to treat cancer?
2. Define the concept of reoxygenation and explain its importance in radiotherapy?
3. Which of the following drugs is an alkylating agent?

    (A) Paclitaxel
    (B) 5-Fluorouracil
    (C) Dacarbazine
    (D) Doxorubicin
    (E) Topotecan

4. Unlike Vincristine, Paclitaxel _____ microtubule formation

    (A) enhances
    (B) inhibits

5. Chemotherapy agents can be classified into cell cycle specific and cell cycle non-specific drugs. Which of the following statements about vinca alkaloids is correct?

   (A) Vinca alkaloids block cell division by inhibiting microtubule function and are G1-phase specific
   (B) Vinca alkaloids which inhibit spindle formation and alignment of chromosomes are M-phase specific
   (C) Vinca alkaloids are cell cycle non-specific drugs as they inhibit spindle formation and alignment of chromosomes
   (D) Vinca alkaloids are most active during the S phase of cell cycle because they exert their cytotoxic activity by inhibiting DNA synthesis
   (E) Vinca alkaloids prevent transcription and replication of DNA and are most active during G1-phase and G2 phase

6. What makes a good therapeutic target?
7. Give an example of a chemotherapeutic drug and a targeted drug. Explain why certain drugs are considered targeted drugs and others are not.
8. A signaling protein inside the cell is mutated and hence constitutively active driving cell proliferation, and resulting in the formation of a tumor. What type of targeted therapy might be effective?

   (A) Monoclonal antibody that prevents growth factors from interacting with the receptor
   (B) Monoclonal antibody that holds the growth factor receptor in the "OFF" position
   (C) Small molecule that selectively binds to the mutated protein
   (D) Monoclonal antibody that selectively binds to the mutated protein

9. What are oncolytic viruses and what is the basis of their specificity for cancer cells?

References

Andtbacka RH et al (2015) Talimogene laherparepvec improves durable response rate in patients with advanced melanoma. J Clin Oncol 33:2780–2788. https://doi.org/10.1200/JCO.2014.58.3377

Benson JD et al (2006) Validating cancer drug targets. Nature 441:451–456. https://doi.org/10.1038/nature04873

Bentzen SM (2006) Preventing or reducing late side effects of radiation therapy: radiobiology meets molecular pathology. Nat Rev Cancer 6:702–713. https://doi.org/10.1038/nrc1950

Blagosklonny MV, Fojo T (1999) Molecular effects of paclitaxel: myths and reality (a critical review). Int J Cancer 83:151–156

Bolus NE (2017) Basic review of radiation biology and terminology. J Nucl Med Technol 45:259–264. https://doi.org/10.2967/jnmt.117.195230

References

Capdeville R, Buchdunger E, Zimmermann J, Matter A (2002) Glivec (STI571, imatinib), a rationally developed, targeted anticancer drug. Nat Rev Drug Discov 1:493–502. https://doi.org/10.1038/nrd839

Chabner BA, Roberts TG Jr (2005) Timeline: chemotherapy and the war on cancer. Nat Rev Cancer 5:65–72. https://doi.org/10.1038/nrc1529

Chapman PB et al (2011) Improved survival with vemurafenib in melanoma with BRAF V600E mutation. N Engl J Med 364:2507–2516. https://doi.org/10.1056/nejmoa1103782

Chatterjee DK, Diagaradjane P, Krishnan S (2011) Nanoparticle-mediated hyperthermia in cancer therapy. Ther Deliv 2:1001–1014

Chen J, Stubbe J (2005) Bleomycins: towards better therapeutics. Nat Rev Cancer 5:102–112. https://doi.org/10.1038/nrc1547

Chiocca EA (2002) Oncolytic viruses. Nat Rev Cancer 2:938–950. https://doi.org/10.1038/nrc948

Corbit KC, Aanstad P, Singla V, Norman AR, Stainier DY, Reiter JF (2005) Vertebrate Smoothened functions at the primary cilium. Nature 437:1018–1021. https://doi.org/10.1038/nature04117

de Sousa Cavalcante L, Monteiro G (2014) Gemcitabine: metabolism and molecular mechanisms of action, sensitivity and chemoresistance in pancreatic cancer. Eur J Pharmacol 741:8–16. https://doi.org/10.1016/j.ejphar.2014.07.041

Delaney G, Jacob S, Featherstone C, Barton M (2005) The role of radiotherapy in cancer treatment: estimating optimal utilization from a review of evidence-based clinical guidelines. Cancer 104:1129–1137. https://doi.org/10.1002/cncr.21324

DeVita VT Jr, Chu E (2008) A history of cancer chemotherapy. Cancer Res 68:8643–8653. https://doi.org/10.1158/0008-5472.can-07-6611

Dick LR, Fleming PE (2010) Building on bortezomib: second-generation proteasome inhibitors as anti-cancer therapy. Drug Discov Today 15:243–249. https://doi.org/10.1016/j.drudis.2010.01.008

Dracham CB, Shankar A, Madan R (2018) Radiation induced secondary malignancies: a review article. Radiat Oncol J 36:85–94. https://doi.org/10.3857/roj.2018.00290

Flaherty KT et al (2012) Combined BRAF and MEK inhibition in melanoma with BRAF V600 mutations. N Engl J Med 367:1694–1703. https://doi.org/10.1056/nejmoa1210093

Fong PC et al (2009) Inhibition of poly(ADP-ribose) polymerase in tumors from BRCA mutation carriers. N Engl J Med 361:123–134. https://doi.org/10.1056/nejmoa0900212

Formenti SC, Demaria S (2009) Systemic effects of local radiotherapy. Lancet Oncol 10:718–726. https://doi.org/10.1016/S1470-2045(09)70082-8

Fu D, Calvo JA, Samson LD (2012) Balancing repair and tolerance of DNA damage caused by alkylating agents. Nat Rev Cancer 12:104–120. https://doi.org/10.1038/nrc3185

Georg D, Thwaites D (2017) Medical physics in radiation Oncology: new challenges, needs and roles. Radiother Oncol 125:375–378. https://doi.org/10.1016/j.radonc.2017.10.035

Gerweck LE (1985) Hyperthermia in cancer therapy: the biological basis and unresolved questions. Cancer Res 45:3408–3414

Glozak MA, Seto E (2007) Histone deacetylases and cancer. Oncogene 26:5420–5432. https://doi.org/10.1038/sj.onc.1210610

Gopal AK et al (2014) PI3K delta inhibition by idelalisib in patients with relapsed indolent lymphoma. N Engl J Med 370:1008–1018. https://doi.org/10.1056/nejmoa1314583

Grant S, Easley C, Kirkpatrick PV (2007) Nature Rev. Drug Discov 6:21–22. https://doi.org/10.1038/nrd2227

Gschwind A, Fischer OM, Ullrich A (2004) The discovery of receptor tyrosine kinases: targets for cancer therapy. Nat Rev Cancer 4:361–370. https://doi.org/10.1038/nrc1360

Hennessy BT, Smith DL, Ram PT, Lu Y, Mills GB (2005) Exploiting the PI3K/AKT pathway for cancer drug discovery. Nat Rev Drug Discov 4:988–1004. https://doi.org/10.1038/nrd1902

Hodi FS et al (2010) Improved survival with ipilimumab in patients with metastatic melanoma. N Engl J Med 363:711–723. https://doi.org/10.1056/NEJMoa1003466

Imming P, Sinning C, Meyer A (2006) Drugs, their targets and the nature and number of drug targets. Nat Rev Drug Discov 5:821–834. https://doi.org/10.1038/nrd2132

Jagtap P, Szabo C (2005) Poly(ADP-ribose) polymerase and the therapeutic effects of its inhibitors. Nat Rev Drug Discov 4:421–440. https://doi.org/10.1038/nrd1718

Jones KL, Buzdar AU (2009) Evolving novel anti-HER2 strategies. Lancet Oncol 10:1179–1187. https://doi.org/10.1016/s1470-2045(09)70315-8

Jordan VC (2003) Tamoxifen: a most unlikely pioneering medicine. Nat Rev Drug Discov 2:205–213. https://doi.org/10.1038/nrd1031

Kamen B (1997) Folate and antifolate pharmacology. Semin Oncol 24:S18-30–S18-39

Kantoff PW et al (2010) Sipuleucel-T immunotherapy for castration-resistant prostate cancer. N Engl J Med 363:411–422. https://doi.org/10.1056/NEJMoa1001294

Kershaw MH, Westwood JA, Darcy PK (2013) Gene-engineered T cells for cancer therapy. Nat Rev Cancer 13:525–541. https://doi.org/10.1038/nrc3565

Khan ZA, Tripathi R, Mishra B (2012) Methotrexate: a detailed review on drug delivery and clinical aspects. Expert Opin Drug Deliv 9:151–169. https://doi.org/10.1517/17425247.2012.642362

Kim JA (2003) Targeted therapies for the treatment of cancer. Am J Surg 186:264–268

Kim YC (2014) EGFR, EGFR TKI, and EMSI: a never-ending story. Transl Lung Cancer Res 3:365–367. https://doi.org/10.3978/j.issn.2218-6751.2014.09.10

Koukourakis G, Kelekis N, Armonis V, Kouloulias V (2009) Brachytherapy for prostate cancer: a systematic review. Adv Urol 327945. https://doi.org/10.1155/2009/327945

Lee TF, Yang J, Wuu CS, Liu A, Fang FM, Yeh SA (2015) Radiation oncology and medical physics. Biomed Res Int 2015:297158. https://doi.org/10.1155/2015/297158

Li F, Jiang T, Li Q, Ling X (2017) Camptothecin (CPT) and its derivatives are known to target topoisomerase I (Top1) as their mechanism of action: did we miss something in CPT analogue molecular targets for treating human disease such as cancer? Am J Cancer Res 7:2350–2394

Longley DB, Harkin DP, Johnston PG (2003) 5-fluorouracil: mechanisms of action and clinical strategies. Nat Rev Cancer 3:330–338. https://doi.org/10.1038/nrc1074

Manasanch EE, Orlowski RZ (2017) Proteasome inhibitors in cancer therapy. Nat Rev Clin Oncol 14:417–433. https://doi.org/10.1038/nrclinonc.2016.206

Markham A (2017) Copanlisib: first global approval. Drugs 77:2057–2062. https://doi.org/10.1007/s40265-017-0838-6

Matson DR, Stukenberg PT (2011) Spindle poisons and cell fate: a tale of two pathways. Mol Interv 11:141–150. https://doi.org/10.1124/mi.11.2.12

Maude SL et al (2018) Tisagenlecleucel in children and young adults with B-Cell lymphoblastic leukemia. N Engl J Med 378:439–448. https://doi.org/10.1056/NEJMoa1709866

Mellstedt H, Niederwieser D, Ludwig H (2008) The challenge of biosimilars. Ann Oncol 19:411–419. https://doi.org/10.1093/annonc/mdm345

Neelapu SS et al (2017) Axicabtagene ciloleucel CAR T-Cell therapy in refractory large B-Cell lymphoma. N Engl J Med 377:2531–2544. https://doi.org/10.1056/NEJMoa1707447

Overholt BF et al (2005) Photodynamic therapy with porfimer sodium for ablation of high-grade dysplasia in Barrett's esophagus: international, partially blinded, randomized phase III trial. Gastrointest Endosc 62:488–498. https://doi.org/10.1016/j.gie.2005.06.047

Palucka K, Banchereau J (2012) Cancer immunotherapy via dendritic cells. Nat Rev Cancer 12:265–277. https://doi.org/10.1038/nrc3258

Pastan I, Hassan R, Fitzgerald DJ, Kreitman RJ (2006) Immunotoxin therapy of cancer. Nat Rev Cancer 6:559–565. https://doi.org/10.1038/nrc1891

Pawlik TM, Keyomarsi K (2004) Role of cell cycle in mediating sensitivity to radiotherapy. Int J Radiat Oncol Biol Phys 59:928–942. https://doi.org/10.1016/j.ijrobp.2004.03.005

Paz-Ares L, Bezares S, Tabernero JM, Castellanos D, Cortes-Funes H (2003) Review of a promising new agent–pemetrexed disodium. Cancer 97:2056–2063. https://doi.org/10.1002/cncr.11279

Pommier Y (2006) Topoisomerase I inhibitors: camptothecins and beyond. Nat Rev Cancer 6:789–802. https://doi.org/10.1038/nrc1977

Postow MA, Callahan MK, Wolchok JD (2015) Immune checkpoint blockade in cancer therapy. J Clin Oncol 33:1974–1982. https://doi.org/10.1200/JCO.2014.59.4358

References

Prise KM, O'Sullivan JM (2009) Radiation-induced bystander signalling in cancer therapy. Nat Rev Cancer 9:351–360. https://doi.org/10.1038/nrc2603

Reedijk J (1999) Why does Cisplatin reach Guanine-n7 with competing s-donor ligands available in the cell? Chem Rev 99:2499–2510

Robert C et al (2015a) Improved overall survival in melanoma with combined dabrafenib and trametinib. N Engl J Med 372:30–39. https://doi.org/10.1056/NEJMoa1412690

Robert C et al (2015b) Nivolumab in previously untreated melanoma without BRAF mutation. N Engl J Med 372:320–330. https://doi.org/10.1056/NEJMoa1412082

Robertson CA, Evans DH, Abrahamse H (2009) Photodynamic therapy (PDT): a short review on cellular mechanisms and cancer research applications for PDT. J Photochem Photobiol B 96:1–8. https://doi.org/10.1016/j.jphotobiol.2009.04.001

Rosenberg SA, Restifo NP, Yang JC, Morgan RA, Dudley ME (2008) Adoptive cell transfer: a clinical path to effective cancer immunotherapy. Nat Rev Cancer 8:299–308. https://doi.org/10.1038/nrc2355

Ruat M, Hoch L, Faure H, Rognan D (2014) Targeting of smoothened for therapeutic gain. Trends Pharmacol Sci 35:237–246. https://doi.org/10.1016/j.tips.2014.03.002

Sabatini DM (2006) mTOR and cancer: insights into a complex relationship. Nat Rev Cancer 6:729–734. https://doi.org/10.1038/nrc1974

Sagar J, Chaib B, Sales K, Winslet M, Seifalian A (2007) Role of stem cells in cancer therapy and cancer stem cells: a review. Cancer Cell Int 7:9. https://doi.org/10.1186/1475-2867-7-9

Sawyers C (2004) Targeted cancer therapy. Nature 432:294–297. https://doi.org/10.1038/nature03095

Schumacher TN, Schreiber RD (2015) Neoantigens in cancer immunotherapy. Science 348:69–74. https://doi.org/10.1126/science.aaa4971

Scott AM, Wolchok JD, Old LJ (2012) Antibody therapy of cancer. Nat Rev Cancer 12:278–287. https://doi.org/10.1038/nrc3236

Sharma P, Allison JP (2015) Immune checkpoint targeting in cancer therapy: toward combination strategies with curative potential. Cell 161:205–214. https://doi.org/10.1016/j.cell.2015.03.030

Shaw AT et al (2013) Crizotinib versus chemotherapy in advanced ALK-positive lung cancer. N Engl J Med 368:2385–2394. https://doi.org/10.1056/NEJMoa1214886

Shewach DS, Kuchta RD (2009) Introduction to cancer chemotherapeutics. Chem Rev 109:2859–2861. https://doi.org/10.1021/cr900208x

Silvestri R (2013) New prospects for vinblastine analogues as anticancer agents. J Med Chem 56:625–627. https://doi.org/10.1021/jm400002j

Strebhardt K, Ullrich A (2008) Paul Ehrlich's magic bullet concept: 100 years of progress. Nat Rev Cancer 8:473–480. https://doi.org/10.1038/nrc2394

Swann JB, Smyth MJ (2007) Immune surveillance of tumors. J Clin Invest 117:1137–1146. https://doi.org/10.1172/JCI31405

Thorn CF, Oshiro C, Marsh S, Hernandez-Boussard T, McLeod H, Klein TE, Altman RB (2011) Doxorubicin pathways: pharmacodynamics and adverse effects. Pharmacogenet Gen 21:440–446. https://doi.org/10.1097/FPC.0b013e32833ffb56

Trott KR (1982) Experimental results and clinical implications of the four R's in fractionated radiotherapy. Radiat Environ Biophys 20:159–170

Trowsdale J, Knight JC (2013) Major histocompatibility complex genomics and human disease. Annu Rev Genomics Hum Genet 14:301–323. https://doi.org/10.1146/annurev-genom-091212-153455

Valabrega G, Montemurro F, Aglietta M (2007) Trastuzumab: mechanism of action, resistance and future perspectives in HER2-overexpressing breast cancer. Ann Oncol 18:977–984. https://doi.org/10.1093/annonc/mdl475

Verma S et al (2012) Trastuzumab emtansine for HER2-positive advanced breast cancer. N Engl J Med 367:1783–1791. https://doi.org/10.1056/nejmoa1209124

Vlashi E, Pajonk F (2015) Cancer stem cells, cancer cell plasticity and radiation therapy. Semin Cancer Biol 31:28–35. https://doi.org/10.1016/j.semcancer.2014.07.001

Voorhees PM, Orlowski RZ (2006) The proteasome and proteasome inhibitors in cancer therapy. Annu Rev Pharmacol Toxicol 46:189–213. https://doi.org/10.1146/annurev.pharmtox.46.120604.141300

Weiner GJ (2010) Rituximab: mechanism of action. Semin Hematol 47:115–123. https://doi.org/10.1053/j.seminhematol.2010.01.011

Wilhelm S et al (2006) Discovery and development of sorafenib: a multikinase inhibitor for treating cancer. Nat Rev Drug Discov 5:835–844. https://doi.org/10.1038/nrd2130

Witzig TE et al (2002) Randomized controlled trial of yttrium-90-labeled ibritumomab tiuxetan radioimmunotherapy versus rituximab immunotherapy for patients with relapsed or refractory low-grade, follicular, or transformed B-cell non-Hodgkin's lymphoma. J Clin Oncol 20:2453–2463. https://doi.org/10.1200/jco.2002.11.076

Zhang J, Yang PL, Gray NS (2009) Targeting cancer with small molecule kinase inhibitors. Nat Rev Cancer 9:28–39. https://doi.org/10.1038/nrc2559

Further Reading

Devita H (2011) Rosenberg's cancer: principles and practice of oncology, 9th edn. Lippincott Williams and Wilkins

DeVita VT Jr, Chu E (2008) A history of cancer chemotherapy. Cancer Res 68:8643–8653. https://doi.org/10.1158/0008-5472.can-07-6611

Holland-Frei cancer medicine, 9th. edn. Wiley-Blackwell (2017) ISBN: 978-1-118-93469-2

Sawyers C (2004) Targeted cancer therapy. Nature 432:294–297. https://doi.org/10.1038/nature03095

Schilsky RL (2010) Personalized medicine in oncology: the future is now. Nat Rev Drug Discovery 9:363–366

Sharma P, Allison JP (2015) Immune checkpoint targeting in cancer therapy: toward combination strategies with curative potential. Cell 161:205–214. https://doi.org/10.1016/j.cell.2015.03.030

Strebhardt K, Ullrich A (2008) Paul Ehrlich's magic bullet concept: 100 years of progress. Nat Rev Cancer 8:473–480. https://doi.org/10.1038/nrc2394

Chapter 3
Cancer Drug Resistance

Resistance to conventional and targeted therapeutics is a fundamental reason for treatment failure in many cancer patients (Gottesman 2002). Drug resistance is the main obstacle to improve the clinical outcome of current cancer therapies (Groenendijk and Bernards 2014). The anti-cancer activity of a drug can be limited by a broad variety of molecular events at different levels of drug action in a **cell autonomous** and **non-cell-autonomous** manner. Resistance can be intrinsic or acquired. **Intrinsic resistance** is the innate capability of a cancer cell to resist therapy due to the presence of pre-existing factors. Conversely, **acquired resistance** refers to the ability of a cancer cell to resist a treatment to which it was previously susceptible. Acquired resistance develops during the treatment and is caused by adaptive responses. Therapy resistance can be caused by host factors or properties or alterations within the cancer cells. In response to therapeutic challenge, tumor cells exploit both genetic and epigenetic evasive mechanisms. **Host factors** such as poor absorption or rapid metabolism limit the amount of systemically administered drugs that reach the tumor. Drugs, in particular monoclonal antibodies and immunotoxins cannot always penetrate sufficiently the tissue of bulky tumors. Adverse effects often lead to the need to treat the tumor with a suboptimal dose. Taken together, these host factors prevent the drug molecules to be at the side of action in an efficient concentration. Furthermore, the non-cancer cells of the host-tumor environment can affect the drug response of the tumor. The mechanisms underlying therapy resistance can also be classified according to the level of interference with the drug action (Fig. 3.1). Resistance to treatment can occur upstream, downstream and at the level of the therapeutic target (Holohan et al. 2013). Mechanisms that operate **upstream of the molecular target** include pharmakinetic host factors leading to drug inactivation or lack of activation and decreased influx and increased efflux of the drug. Alterations of the **drug target itself** such as changes in its level of expression or mutations can render the target insensitive to treatment. The reactivation of the pathway via downstream components or the activation of redundant and alternative signal transduction pathways, enhanced prosurvival or dysfunctional apoptosis signaling, and efficient DNA damage repair represent mechanisms the confer therapy resistance operating downstream of the

© Springer Nature Switzerland AG 2019
W. Link, *Principles of Cancer Treatment and Anticancer Drug Development*,
https://doi.org/10.1007/978-3-030-18722-4_3

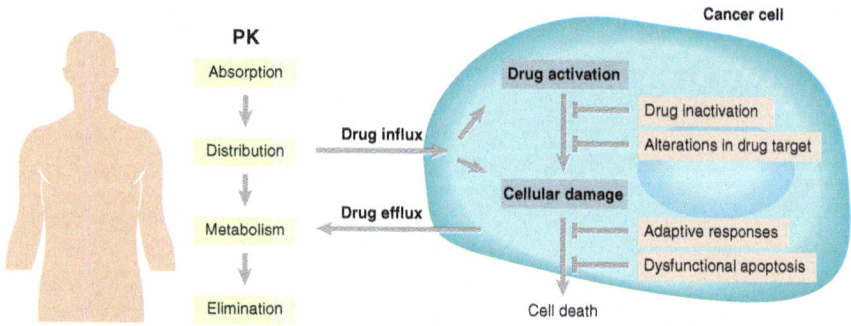

Fig. 3.1 Mechanisms of drug resistance. Cancer cells can escape the effect of treatment by numerous mechanisms which can be classified into mechanisms that operate upstream of the molecular, downstream or on the drug target itself. Adapted from Nat Rev Cancer 13, 714

drug target. It is important to note that many of these mechanisms are relevant for cytotoxic chemotherapy as well as for targeted therapies.

3.1 Mechanisms Upstream of the Molecular Target

Increased Expression of Drug Transporter
Excessive efflux of drugs from the target cells is a notorious mechanism of therapy resistance. The **ATP-binding cassette (ABC) transporters** are a family of transmembrane proteins responsible for the flux of structurally and mechanistically unrelated anti-cancer drugs across the plasma membrane (Fig. 3.2; Dean et al. 2001).

Chemotherapy resistant tumors hijacking this mechanism and actively expel the cytotoxic drugs from cells, maintaining the drug concentration within the cells below the toxic level. The best studied family member is the membrane-bound glycoprotein **multi-drug resistance protein 1** (MDR1) (Borst et al. 1999). MDR1 is expressed in many tissues at low level but has been identified to be overexpressed in many tumors and its expression correlates with poor response to several chemotherapies. ABC transporters promote the efflux of several hydrophobic compounds conferring resistance to antimetabolites, taxanes, and topoisomerase inhibitors. Targeted drugs including imatinib, erlotinib, sunitinib and nilotinib have also been suggested to be substrates of efflux mediated by ABC transporters.

Reduced Drug Uptake
Several drugs need to be actively transported into the cell via specific membranes receptors (Holohan et al. 2013). Cancer cells can eliminate of modify these cell surface molecules and thereby inhibit or reduce the drug concentration inside the cell. Folic acid analogues such as methotrexate require the expression of folate binding protein or the reduced folate transporter for the cell influx (van der Heijden et al.

3.1 Mechanisms Upstream of the Molecular Target

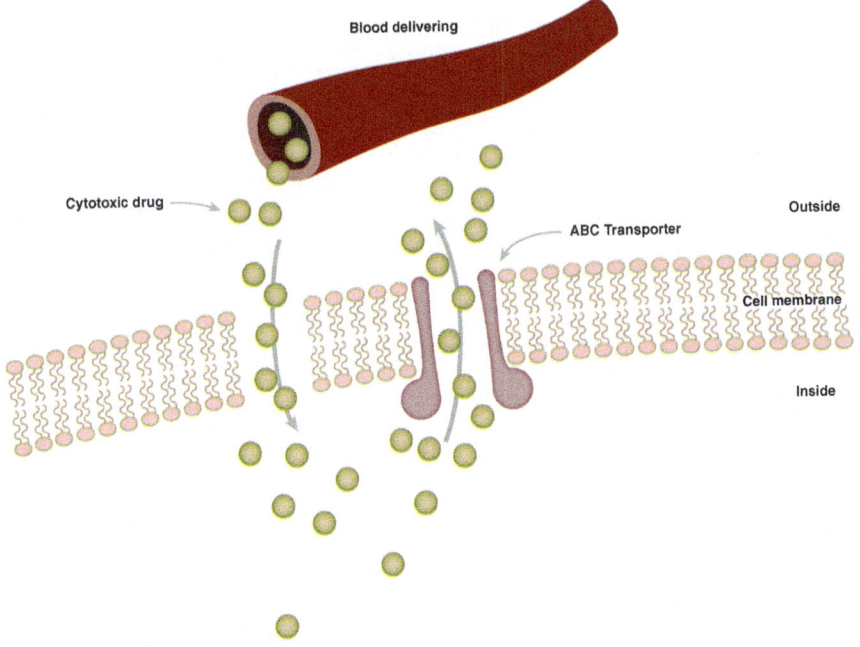

Fig. 3.2 ATP-binding cassette (ABC) transporters confer resistance to structurally and mechanistically unrelated anti-cancer drugs by pumping out the cytotoxic drugs from cells maintaining the intracellular drug concentration low

2007). Accordingly, resistance to these drugs often occur by mutations within the transporter proteins. In a similar fashion, somatic mutations and possibly also germ line variations in nucleoside transporters prevent nucleoside analogs including 5-fluorouracil (5FU) and gemcitabine from entering the target cells and exert their anti-cancer action representing a major challenge to the management of many solid cancers such as colorectal and pancreatic cancer.

Drug Inactivation
Drug metabolism and detoxification are key mechanisms of therapy resistance in many cancer types. A drug molecule might be metabolized by specific enzymes in the host before it reaches the target tissue. Alternatively, the drug can be modified and thereby inactivated locally in the environment of the tumor or within the tumor cells. As most drugs are metabolized by cytochrome P450 enzymes which are known to be highly polymorphic, there are large pharmacokinetic differences between patients (Rochat 2005). For example, interindividual differences in CYP2D6, CYP2C19 and CYP2B6 due to polymorphic allels determine the capacity to metabolize cyclophosphamide and tamoxifen and influence the therapeutic effect of these drugs. Similar to cytochrome P450 enzymes, gluthathione *S*-transferase an enzyme that catalyzes the conjugation of the reduced form of glutathione can modify and thereby detoxify anti-cancer drug molecules. Drugs known to be inactivated by glutathione include

alkylating and alkylating-like agents and increased activity of glutathione-*S*-transferase in some cancers has been reported to confer resistance to these drugs (Townsend and Tew 2003).

Lack of Prodrug Activation
Many anticancer drugs are prodrugs and require metabolic activation to exert their therapeutic effect (Aghi et al. 2000). As mentioned above many of the chemotherapeutic antimetabolites are modified by kinases and phosphoribosyl transferases to generate their active nucleosides. Acute myelogenous leukemias can become resistant to the treatment with Cytarabine by inactivation of enzymes involved in the metabolic activation of this prodrug. In order to become active Cytarabine needs to be converted into cytarabine-monophosphate and subsequently into cytarabine triphosphate via two phosphorylation steps. Mutation or reduced expression of one of the responsible kinases, namely deoxycytidine kinase has been reported to confer resistance to Cytarabine treatment. Similarly, the alkylating agents cyclophosphamide and ifosfamide are also prodrugs that require metabolic activation in the liver by Cyp 450 enzymes. Accordingly, the expression level and enzymatic activity of these enzymes determine the therapeutic effect of these drugs (Rochat 2005).

3.2 Mechanisms at the Level of the Molecular Target

Drug Target Modification
Mechanisms underlying therapy resistance can also operate directly on the molecular target of the drug (Holohan et al. 2013). The therapeutic target can suffer changes in the expression level or mutate and in turn reduce responsiveness to anticancer drugs (Fig. 3.3). The expression of thymidylate synthase, the molecular target of several antimetabolites including Pemetrexed is regulated by a negative feedback loop at the post-translational level. As a consequence, inhibition of thymidylate synthase inhibits the translation of thymidylate synthase transcripts (Ozasa et al. 2010). In this model, the anti-cancer drug downregulates the expression of its target and attenuates the drug response. Similarly, Doxorubicin has been reported to downregulate its target topoisomerase IIα. A very common mechanism that reduces the response to kinase inhibitors are gatekeeper mutations within the catalytic domain of oncogenic kinases. Secondary EGFR-T790M gatekeeper mutation have been reported to confer resistance to the treatment of non-small-cell lung cancer with gefitinib and erlotinib leading to relapse within less than one year. Similarly, resistance to Crizotinib is commonly due to mutations in the ALK tyrosine kinase domain or ALK fusion gene amplifications and a mutation in the gatekeeper residue of BCR/ABL T315 has been shown to prevent Imatinib from binding to the kinase while preserving the catalytic activity (O'Hare et al. 2007). This mutation is one of several alterations in BCR/ABL known to confer resistance to Imatinib.

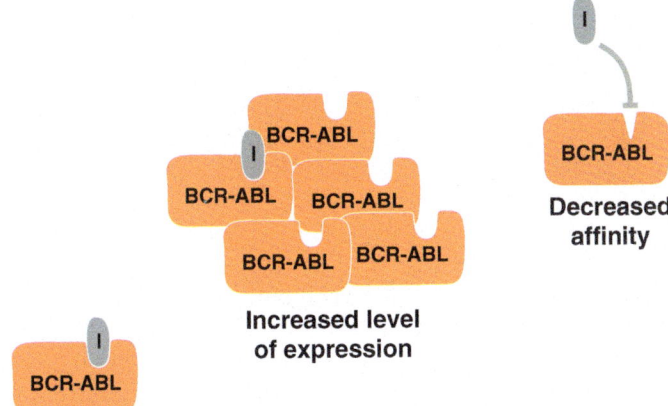

Fig. 3.3 Two mechanisms by which the cancer cells can evade the anti-cancer effect of the kinase inhibitor Imatinib (I, gray). The cell might increase the level of expression of the molecular target BCR-ABL (orange). Another frequent strategy to evade Imatinib effect is to acquire a mutation that leads to a decreased affinity of the binding of Imatinib to the kinase domain of its target. Adapted from Scientific World Journal 6, 918

3.3 Mechanisms Downstream of the Molecular Target

Pathway Reactivation
Therapy resistance can take place through pathway reactivation either via an intrinsic feedback loop or the acquired activation of a downstream component of the targeted pathway (Holohan et al. 2013). PI3K inhibitors reduce the generation of the second messenger PIP3 and in turn downregulate the activity of the oncokinase AKT. However, the effect of PI3K inhibitors is limited by components of a downstream branch which reactivate the pathway. Reduced AKT activity leads to the inhibition of mTOR and in turn S6 kinase (S6K) and eventually to the reactivation of PI3K/AKT signaling via enhanced function of the S6K substrate insulin receptor substrate (IRS) (Manning and Cantley 2007). Similarly, the pseudokinase TRIB2 has been reported to activate PI3K/AKT signaling even in the presence of PI3K inhibitors by direct interaction of TRIB2 with AKT leading to increased enzymatic activity (Hill et al. 2017). The appearance of resistance to BRAF inhibitors in melanoma patients is mostly due to the restoration of MAPK signalling by BRAF independent activation of MEK, a downstream component of the pathway (Sun et al. 2014). Several mechanisms to circumvent BRAF in MEK activation have been described including loss of the GTPase-activating protein NF1, mutation in NRAS or its overexpression, CRAF overexpression and mutation or overexpression of MEK itself (Ferreira et al. 2017).

Activation of Alternative Signaling Pathways
A common mechanism of drug resistance to targeted drugs works through the activation of parallel or redundant signalling pathways (Holohan et al. 2013). The clinical

outcome of BRAF treatment in melanoma patients is limited by the activation of the PI3K/AKT pathway through the receptor tyrosine kinases MET and IGFR. In addition, mutations leading to the loss of function of the lipid phosphatase PTEN which antagonizes the activity of PI3K can activate PI3K/AKT signalling (Paraiso et al. 2011). Similarly, activating mutations in AKT have been reported to confer resistance to BRAF inhibitors. The activation of PI3K/AKT signalling pathway is capable of maintaining survival of melanoma cells in the presence of BRAF inhibitors.

DNA Damage Repair

Many chemotherapeutic drugs act against cancer cells by damaging their DNA. Therefore, alterations that enable the target cell to increase its capacity to repair the damaged DNA represent an obvious drug resistance mechanism (Lord and Ashworth 2012). Hypermethylation MLH1, a gene involved in mismatch repair (MMR), leads to the silencing of the gene and has been identified as a powerful mechanism of resistance to platinum based compounds like Cisplatin and Carboplatin. Similarly, the repair protein methylguanine methyltransferase or MMR deficiency confers resistance of glioblastoma multiforme to Temozolomide, the only approved chemotherapeutic drug to treat this devastating disease. Temozolomide methylates purine bases of DNA, but the enzyme methylguanine methyltransferase (MGMT) can reverse methylation of the O6 position of O6-methylguanine which represents the primary cytotoxic lesion produced by Temozolomide (Lee 2016). Overexpression of MGMT results in increased repair of Temozolomide-induced DNA damage and in turn drug resistance. Conversely, the methylation of the MGMT promoter reduces MGMT expression and is correlated with efficiency of Temozolomide. Accordingly, methylation status of the MGMT promoter is currently the most relevant genetic fingerprint to predict clinical response to Temozolomide treatment. Another system required to repair DNA damaged by many chemotherapeutic agents including Cisplatin is nucleotide-excision repair (NER). Interestingly, testicular cancer known to be very vulnerable to Cisplatin expresses very low level of the excision repair cross-complementing 1 (ERCC1) which is an essential component of the NER system.

Reduced Susceptibility to Apoptosis

The vast majority of anti-cancer drugs kill cancer cells by promoting apoptosis. Accordingly, a major downstream mechanism of resistance to many of these treatments involves the deregulation of apoptosis pathways by mutations or changes in the expression of key apoptosis proteins including, Bcl-2, Bcl-XL, BIM, FLIP and inhibitor of apoptosis proteins (IAPs) (Longley and Johnston 2005). BIM has been reported to play an important role in apoptosis mediated by several anti-cancer drugs including the tryrosine kinase inhibitors Gefitinib, Erlotinib and Imatinib. The expression of the pro-apoptotic gene BIM is regulated by FOXO transcription factors which are activated in the absence of survival signalling through AKT. Accordingly, BIM expression correlates with clinical response to tyrosine kinase inhibitors and PI3K inhibitors and germline deletion of the gene has been associated with intrinsic resistance to these kinase inhibitors.

3.3 Mechanisms Downstream of the Molecular Target

Tumor Heterogeneity and Epithelial-to-Mesenchymal Transition

The heterogeneity of tumors is a key driver of drug resistance leading to therapeutic failure (Turner and Reis-Filho 2012). Intra-tumor heterogeneity comprises both genetic and epigenetic components distinct from the founding immortalized cell. Tumor heterogeneity supports a view of cancer as an abnormal tissue comprising a complex interplay between tumor cells and the normal cellular counterparts in the organ in which they reside. The ability to adapt and transition between different cellular phenotypes will in turn increase the likelihood of tumor cell survival. An important example of cellular phenotypic plasticity is the **epithelial-to-mesenchymal transition** (EMT) (Henriques et al. 2018). EMT is an evolutionary conserved process by which epithelial cells reversibly abandon their characteristic cell polarity and cell-cell adhesions, in favour of migratory and invasive properties typical of mesenchymal (fibroblast-like) cells during embryonic development and wound healing. In cancer, EMT is associated with resistance to chemotherapeutics, immune evasion, metastasis and poor clinical outcome. It is thought that transcription factors that drive EMT including slug confer resistance to EGFR in Non-small-cell lung carcinoma through the repression of BIM expression and increased caspase-9 activity in NSCLC. Recently, it has been shown that transcriptional profiles of tumors resistant to immunotherapy with immune checkpoint inhibitors resemble those induced by EMT.

Tumor Microenvironment

The tumor microenvironment comprises numerous normal cell types, **extracellular matrix** (ECM) components, and soluble growth factors. It is increasingly appreciated that a dynamic tumor microenvironment facilitates many steps in the tumor development including cancer initiation, growth and metastasis (Lu et al. 2012). Novel combinations of biochemical and biophysical signals from the cell-cell and cell-ECM interactions in the tumor microenvironment regulate responsiveness to growth factors that determine cell behaviors and affect cancer therapy responses (Tredan et al. 2007). The tumor microenvironment also determines therapeutic efficacy. Many soluble factors, including growth factors, hormones and cytokines (including chemotactic cytokines; chemokines) are secreted by tumor cells and other cells in the tumor microenvironment such as **cancer-associated fibroblasts** (CAFs) or **tumor-associated macrophages** (TAMs). For example, resistance of BRAF mutant melanoma cells to BRAF inhibitors has been shown to be mediated by the hepatocyte growth factor (HGF) released by non-cancerous tumor-associated stromal cells (Straussman et al. 2012).

Thought Questions

1. Describe three alternative mechanisms of drug resistance and provide an example for each.
2. What is the predominant mechanism of resistance to Temozolomide (TMZ)
   - (A) Alteration of hydrofolate reductase activity (DHFR)
   - (B) Decreased formation of polyglutamate
   - (C) Overexpression of methylguanine methyltransferase (MGMT)
   - (D) Increased levels of PTEN
   - (E) Overexpression multi-drug resistance protein 1 (MDR1).

References

Aghi M, Hochberg F, Breakefield XO (2000) Prodrug activation enzymes in cancer gene therapy. J Gene Med 2:148–164. https://doi.org/10.1002/(SICI)1521-2254(200005/06)2:3%3c148:AID-JGM105%3e3.0.CO;2-Q

Borst P, Evers R, Kool M, Wijnholds J (1999) The multidrug resistance protein family. Biochim Biophys Acta 1461:347–357

Dean M, Hamon Y, Chimini G (2001) The human ATP-binding cassette (ABC) transporter superfamily. J Lipid Res 42:1007–1017

Ferreira BI, Lie MK, Engelsen AST, Machado S, Link W, Lorens JB (2017) Adaptive mechanisms of resistance to anti-neoplastic agents. Medchemcomm 8:53–66. https://doi.org/10.1039/c6md00394j

Gottesman MM (2002) Mechanisms of cancer drug resistance. Annu Rev Med 53:615–627. https://doi.org/10.1146/annurev.med.53.082901.103929

Groenendijk FH, Bernards R (2014) Drug resistance to targeted therapies: Deja vu all over again. Mol Oncol 8:1067–1083. https://doi.org/10.1016/j.molonc.2014.05.004

Henriques V, Martins T, Link W, Ferreira BI (2018) The emerging therapeutic landscape of advanced melanoma. Curr Pharm Des 24:549–558. https://doi.org/10.2174/1381612824666180125093357

Hill R et al (2017) TRIB2 confers resistance to anti-cancer therapy by activating the serine/threonine protein kinase AKT. Nat Commun 8:14687. https://doi.org/10.1038/ncomms14687

Holohan C, Van Schaeybroeck S, Longley DB, Johnston PG (2013) Cancer drug resistance: an evolving paradigm. Nat Rev Cancer 13:714–726. https://doi.org/10.1038/nrc3599

Lee SY (2016) Temozolomide resistance in glioblastoma multiforme. Genes Dis 3:198–210. https://doi.org/10.1016/j.gendis.2016.04.007

Longley DB, Johnston PG (2005) Molecular mechanisms of drug resistance. J Pathol 205:275–292. https://doi.org/10.1002/path.1706

Lord CJ, Ashworth A (2012) The DNA damage response and cancer therapy. Nature 481:287–294. https://doi.org/10.1038/nature10760

Lu P, Weaver VM, Werb Z (2012) The extracellular matrix: a dynamic niche in cancer progression. J Cell Biol 196:395–406. https://doi.org/10.1083/jcb.201102147

Manning BD, Cantley LC (2007) AKT/PKB signaling: navigating downstream Cell 129:1261–1274. https://doi.org/10.1016/j.cell.2007.06.009

O'Hare T, Eide CA, Deininger MW (2007) Bcr-Abl kinase domain mutations, drug resistance, and the road to a cure for chronic myeloid leukemia. Blood 110:2242–2249. https://doi.org/10.1182/blood-2007-03-066936

References

Ozasa H, Oguri T, Uemura T, Miyazaki M, Maeno K, Sato S, Ueda R (2010) Significance of thymidylate synthase for resistance to pemetrexed in lung cancer. Cancer Sci 101:161–166. https://doi.org/10.1111/j.1349-7006.2009.01358.x

Paraiso KH et al (2011) PTEN loss confers BRAF inhibitor resistance to melanoma cells through the suppression of BIM expression. Cancer Res 71:2750–2760. https://doi.org/10.1158/0008-5472.CAN-10-2954

Rochat B (2005) Role of cytochrome P450 activity in the fate of anticancer agents and in drug resistance: focus on tamoxifen, paclitaxel and imatinib metabolism. Clin Pharmacokinet 44:349–366. https://doi.org/10.2165/00003088-200544040-00002

Straussman R et al (2012) Tumour micro-environment elicits innate resistance to RAF inhibitors through HGF secretion. Nature 487:500–504. https://doi.org/10.1038/nature11183

Sun C et al (2014) Reversible and adaptive resistance to BRAF(V600E) inhibition in melanoma. Nature 508:118–122. https://doi.org/10.1038/nature13121

Townsend DM, Tew KD (2003) The role of glutathione-S-transferase in anti-cancer drug resistance. Oncogene 22:7369–7375. https://doi.org/10.1038/sj.onc.1206940

Tredan O, Galmarini CM, Patel K, Tannock IF (2007) Drug resistance and the solid tumor microenvironment. J Natl Cancer Inst 99:1441–1454. https://doi.org/10.1093/jnci/djm135

Turner NC, Reis-Filho JS (2012) Genetic heterogeneity and cancer drug resistance. Lancet Oncol 13:e178–e185. https://doi.org/10.1016/S1470-2045(11)70335-7

van der Heijden JW, Dijkmans BA, Scheper RJ, Jansen G (2007) Drug Insight: resistance to methotrexate and other disease-modifying antirheumatic drugs–from bench to bedside. Nat Clin Pract Rheumatol 3:26–34. https://doi.org/10.1038/ncprheum0380

Further Reading

Groenendijk FH, Bernards R (2014) Drug resistance to targeted therapies: Deja vu all over again. Mol Oncol 8:1067–1083. https://doi.org/10.1016/j.molonc.2014.05.004

Holland-Frei cancer medicine, 9th. edn. Wiley-Blackwell (2017) ISBN: 978-1-118-93469-2

Holohan C, Van Schaeybroeck S, Longley DB, Johnston PG (2013) Cancer drug resistance: an evolving paradigm. Nat Rev Cancer 13:714–726. https://doi.org/10.1038/nrc3599

Shibue T, Weinberg RA (2017) EMT, CSCs, and drug resistance: the mechanistic link and clinical implications. Nat Rev Clin Oncol 14:611–629

Chapter 4
Drug Discovery and Development

The Paradigm Shift from Empiric to Knowledge-Based Drug Discovery
In the past, finding new medicines to treat human diseases was based on trial and error without pre-existing knowledge about the underlying mechanism of the therapeutic effect (Drews 2000). Understanding the molecular mechanism of action of these empirically identified and sometimes optimized drugs involved a tedious research process often during decades or even centuries (Ferreira et al. 2015). Despite the important impact of drugs like morphine and aspirin in clinical practice, their mode of action and therapeutic targets remained unknown for a long time (Box 1) (Sneader 2000).

> **Box 1**
> **Short historical perspective of drug discovery**
> Ancient civilizations identified the beneficial effects of plant extracts for different disease indications. This approach was based on trial and error using mainly water and water/alcohol-based plant extracts. Advances in organic chemistry and chemical analysis has enabled the isolation and chemical characterization of the active components within crude extracts, for example, the isolation of morphine from opium through crystallization between 1803 and 1805 that is thought to be the first active principle isolated from a plant source. Despite his limited scientific knowledge, Sertürner recognized the importance of experiments to observe the effects of morphine in vivo and conducted rudimentary "preclinical" and "clinical" studies. Strikingly (and despite its important impact in clinical practice), the molecular mode of action of morphine remained unknown for over 150 years and it was not until the early 1950s when the existence of specific receptors for morphine and its derivatives were proposed 9. In the 1970s the existence of these receptors were confirmed and cloned 10. Another good example of drug discovery in the past is Aspirin. Ancient civilizations used willow bark for centuries while in the 19th century,

many chemists tried to purify the active compound from willow bark. Willow bark contains salicin that is metabolized in our body to the active agent salicylic acid. To identify derivatives of salicylic acid, acetylsalicylic acid, the first synthetic drug was synthesized in 1897 by Felix Hoffman and Arthur Eichengrün 11 and marketed in 1899 as aspirin. Remarkably the molecular target of aspirin was only identified in the 1970s (almost 100 years after its synthesis) when J. R. Vane demonstrated that aspirin irreversibly inhibits the enzyme cyclooxygenase (COX) that catalyzes the synthesis of prostaglandins. At the beginning of the 20th century the German physician Paul Ehrlich set the stage for modern anti-cancer therapy with his magic bullet concept which turned out to be a particularly inspiring and influential idea. Ehrlich developed a receptor theory (side-chain theory) postulating the existence of specific receptors for toxins, drugs and other substances. Accordingly, it was hypothesised that diseases should be treatable with chemical compounds that interact with disease-specific cellular components with limited effects on healthy tissues.

Similarly, the first generation of anti-cancer drugs have been identified through serendipity and then optimized on the basis of cytotoxicity in growth proliferation models. Based on technological progress within life sciences which allowed for the knowledge and understanding of the molecular bases of many human pathologies, our way to find new medicines has dramatically changed during the last few decades. Modern approaches to develop new medicines invert the steps of the discovery process (Fig. 4.1; Ferreira et al. 2015).

This new framework is based on understanding the disease at the molecular level. Whereas, traditionally drug development was empirical, nowadays the way we find and develop new medicines is knowledge-based. The process starts with the identification and validation of a molecular target followed by the discovery of target-specific agents. The availability of the complete human genome sequence and our advanced understanding of the molecular basis of disease have provided a plethora of cellular components that represent potential therapeutic targets (Santos et al. 2017). Once identified, an extensive target validation process is required addressing a disease-causative effect of the target and decisively, the therapeutic potential of this candidate

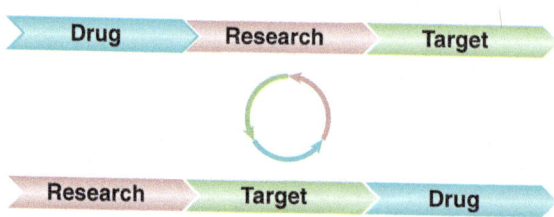

Fig. 4.1 The way we find new medicines have changed during the last decades. In the past, a drug was found before its molecular target was identified. Modern drug discovery is based on the molecular understanding of the disease and starts out with a validated molecular target

4 Drug Discovery and Development

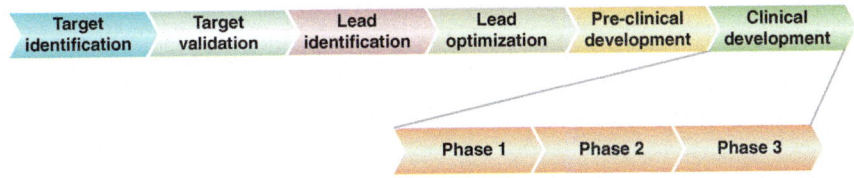

Fig. 4.2 The process of modern drug discovery and development can be divided in several different steps. The discovery process leads to a lead compound. Drug development is the process of bringing a new drug to the market and includes preclinical and clinical development. The discovery process usually begins with the discovery of a target which has to be validated before a therapeutic agent can be discovered. The first active agents will be optimized (lead optimization) until they a mature enough to enter preclinical testing. Eventually, the drug candidates have to undergo three phases of evaluation in human patients (clinical development) before they can be approved for their comercialization

target. Modern drugs rely on the previous identification and validation of **disease-relevant targets** which represents the most innovative step within the whole process as it facilitates the development of drugs with new modes of action (Benson et al. 2006). The actual process of drug discovery starts with the availability of a validated target. Once an inhibitor of the selected target has been identified, it has to be modified to optimize its properties. The optimized product is then tested in vitro and in vivo assays to evaluate the pharmacodynamic and pharmacokinetic properties. This phase of the development process is aimed at selecting the most efficient and safest drug candidate for its evaluation in clinical trials. A successful drug candidate has to overcome three phases of clinical testing before it can be approved and enter the market. Nowadays, drug discovery and drug development is a long and expensive process. It takes an average of 12 years and costs about 800 million US dollars to get a new drug from the laboratory to the pharmacy shelf. The process consists of several sequential steps: (1) Target identification, (2) Target validation, (3) Lead identification Identification, (4) Lead optimization, (5) Pre-clinical development and (6) Clinical development (Fig. 4.2; Ferreira et al. 2015).

4.1 Target Identification

The important progress in the molecular understanding of cancer which has been made during the last three decades has profoundly transformed the way we identify and develop anticancer drug. The identification and validation of disease relevant targets are crucial for the development of molecularly targeted anticancer therapies. However, without a thorough understanding of the molecular events driving tumor formation and progression it is difficult to identify therapeutically useful targets (Ferreira et al. 2015). Therefore, these targets often emerge from basic research laboratories dedicated to understand and manipulate the driving molecular events of tumor formation and progression (Knowles and Gromo 2003). Scientists tackle the problem at different molecular levels. They analyze which genes are affected in the disease condition, how that affects the function of the proteins they encode and if it affects how these proteins then interact in living cells and the possible impact on a specific tissue and eventually on a whole organism, in particular, a human patient. Conversely, therapeutic targets might be identified through reverse target validation, a process usually called **target deconvolution** (Box 2) based on an interesting biological activity of a small molecule compound (Terstappen et al. 2007).

Box 2

Target deconvolution

Target deconvolution is the process of identifying the molecular target of a hit compound from a phenotypic screening (assays with a phenotypic readout such as cell viability, cell migration, autophagy, intracellular protein translocation etc., that do not allow to directly infer the molecular target and the mode of action of the active compound) or any compound with a therapeutically relevant biological activity and without a known molecular target. Recent advances in 'omics' technologies, activity profiling and label free methods as well as in silico modelling enable efficient target deconvolution. Target prediction of small molecules using in silico tools is based on structural similarities between an active hit and well-characterized drugs in databases. In silico identified target candidates have to be confirmed by additional experimental procedures. Transcriptomic analysis can reveal similarities to transcriptional profiles produced by compounds with known mode of action using connectivity map algorithms. If more specific information on possible molecular targets or target classes is available, focussed target profiling can be performed. For example, if it is known that the biological process of interest is governed by kinases, the hit compound can be tested in broad panels of kinases (up to 500) to test its capacity to inhibit their activity. Nowadays, chemical proteomics approaches including label free technologies, affinity purification or click chemistry are widely used in target deconvolution efforts. Label-free approaches rely on changes in thermodynamic stability as the result of a protein-drug interaction which can be quantified by different methods. Affinity purification involves immobilization of the active compound onto a solid support to isolate bound protein targets which can be separated by gel electrophoresis and analysed by mass spectrometry. Click chemistry is a relatively new approach and refers to the versatile use of simple chemical reactions and can be applied to develop bifunctional molecules in which the active compound can be labelled with a fluorescent moiety.

Accordingly, a broad variety of research methods, models and technologies are used including genomics, proteomics, cellomics, metabolomics based on biochemical, cellular and animal models. Several different types of targets can be

4.1 Target Identification

Table 4.1 List of target types, examples of therapeutic targets for approved drugs, preferential expression of the drug target and approved drug targeting the corresponding target

Target type	Target	Target expression	Drug
Addiction	BCR-Abl	Cancer cells	Imatinib
Lethality	PARP	Cancer cells	Olaparib
Lineage	CD20	Normal and cancer cells	Rituximab
Microenvironment	VEGFR2	Normal cells	Ramucirumab

distinguished according to their specific expression and strategy of their inhibition (Table 4.1).

Typically, gene sequence and copy number data as well as gene expression profiles obtained from tumor tissues compared to normal tissues enable researches to identify mutated and differentially expressed genes. Recent technological progress enabled the interrogation of many genes in multiple independent samples even at single cell level to identify genetic changes associated with cancer. The conceptual framework of therapeutic targets in oncology is based on the assumption that cancer is caused by stable alterations which are passed on upon sequential cell divisions. Therapeutic success relies on the dependence of cancer cell survival on such changes. A tumor might accumulate genetic and epigenetic alterations and loose dependence on earlier changes. Conversely, continuous dependence of cancer cells upon a specific alteration has been coined **oncogene addiction** (Weinstein and Joe 2008). Therefore, an ideal molecular target for an anticancer drug is specific for the cancer cell and not present in normal cells and essential for cancer cell survival and proliferation. The clinical success of Imatinib proofed this concept to be suitable as it showed that Philadelphia-chromosome positive Chronic Myelogenous Leukaemia (CML) cells seem to be addicted to the continuous activity of oncogenic BRC-Abl and die when its kinase activity is blocked. However, there are many alterations present in cancer cells, which individually do not have any phenotypic impact but the combination of two or three alterations might produce deleterious phenotypes. These combinations of genetic chances present vulnerabilities of cancer cells and might be exploited therapeutically. This concept is known as **synthetic lethality** and can be applied to therapy considering pharmaceutical inhibition of a target as a loss of function change which can destroy a cancer cells in combination with mutant genes while leaving normal cells intact (Kaelin 2005). PARP inhibitors induce a loss of function of Poly ADP ribose polymerases involved in specific DNA repair systems and act as lethal agents only in combination with an additional defect in another DNA repair mechanism. Normal cells with intact DNA repair systems have back up mechanism to manage DNA damage and are less sensitive to PARP inhibitors.

As mentioned above, most cancer cells maintain many features of the cells from which they were derived. Their survival and proliferation depend on similar factors than normal cells of the same cell lineage, a feature that might be exploited therapeutically. This concept is known as **lineage addiction** (Garraway and Sellers 2006). Therapeutic mAb or antibody-drug conjugates such as Rituximab or Rituximab and hyaluronidase human, respectively are directed against the surface antigen CD20 which is specific for cells of the B-cell lineage with a certain differentiation state. These drugs destroy cells of the same lineage expressing CD20 regardless if they are cancer cells or not. Some drugs used for the treatment of cancer target proteins which are not even expressed in cancer cells, but in non-cancerous cells of the **tumor microenvironment**. Tumor might even become addicted to certain support such as growth factors provided by the microenvironment and therapeutic targeting of the support system is considered as a promising strategy to starve and destroy the tumor. The mAb Ramucirumab and the small molecule drugs Sorafenib and Sunitinib target the VEGF receptor mainly expressed by endothelial cells in the environment of the tumor and responsive to growth factors released by the tumor to induce angiogenesis. Blocking the receptor interrupts the stimulatory interaction between tumor and the non-malignant environment.

Target discovery is the identification of molecular vulnerabilities of specific cancers (Lindsay 2003). However, as outlined earlier (see Box 5 in Chap. 2), druggability is also an important aspect to select a therapeutic target. Ideal targets are very rare in cancer cells. Cancer research aims to identify targets that are to some degree essential and specific to cancer cells versus normal cells and their function can be modulated by drugs.

4.2 Target Validation

Target validation is the process of establishing a disease-causative effect and the therapeutic potential of a macromolecule (Benson et al. 2006). Innovative drug discovery needs well-validated targets. Therefore, target identification/validation is the single most important step in the innovation value chain. Poor target validation status is one of the major reasons for drug attrition (Waring et al. 2015). Remarkably, the pharmaceutical industry managed to reduce drug attrition rate due to pharmacokinetic reasons but was much less successful in reducing attrition due to efficacy and safety, two pharmacodynamic factors that depend on the drug targets. Hence, many drug candidates fail in clinical trials because they don't work as expected in human. This observation highlights the importance of thorough target validation and indicates that a therapeutic target is never fully validated until the drug has been tested in humans. A cancer drug target is only truly validated by demonstrating that a given therapeutic agent is clinically effective and acts through the target against which it was designed. The characterization of the disease-relevance and therapeutic potential of a target is a lengthy and complex process. Most candidate targets emerge from basic research as correlated with cancer. Mutations, epigenetic modifications, copy number, gene expression level, protein level, protein modification, protein localization might correlate with cancer cells versus normal cells. However, a correlation does not establish causation. A striking example that illustrates the difference between correlation and causation is the fireman paradox:

4.2 Target Validation

Firemen are often found at burning houses.
Firemen are usually not found at normal houses.
Therefore, firemen cause house fires.
Therefore, prevent fires eliminating firemen.

Similarly, a protein that is consistently overexpressed in cancer cells compared to normal cells might be involved in a defense mechanism against tumorigenesis or the overexpression is completely coincidental. But how can you determine if a molecular target is disease-relevant and has therapeutic potential? The significance of a target in cancer can be determined by genetic manipulation of cells or animal models. Target validation is based on the use of experimental system suitable for modeling the human disease. Each of these approaches presents specific advantages and limitations. Though cancer associated gene mutations or expression level might be a good starting point to establish a drug discovery project, it is impossible to predict the clinical efficacy of an inhibitor based on this data. Further effort will be required to understand the function of the target and to evaluate the impact of its manipulation. After thorough analysis of genetic and protein data available in public and proprietary data bases such as The Cancer Genome Atlas (TCGA) or The Human Protein Atlas, suitable model systems to functionally characterize the role of the target in tumorigenesis will be selected (Cancer Genome Atlas Research et al. 2013).

Cell Based Systems

As cells are relatively cheap and easy to handle, early stage validation often starts with cell based assays. An advantage over animal models (see below) is that cell-based assay using human cells are human models. An important aspect of selecting suitable cell systems is the cancer type in which a differential expression or mutations have been identified. In addition, the endogenous expression level of target transcripts and proteins should be taken into account when choosing cell lines for target overexpression and target knockdown. One of the first steps in the target validation process is to analyze cell phenotypes in the presence and absence of the selected target (Table 4.2).

Ectopic overexpression of the wildtype or mutant gene in a suitable cell line or in non-transformed cells are usually performed in the context of low expression level of the endogenous protein. Conversely, gene silencing using RNA-interference (RNAi) approach can be performed in a transient fashion via short-interfering RNA (siRNA) or stably using short-hairpin RNA (shRNA) in cells with significant target expression. Transient or stable knockdown of the selected target is considered as a surrogate model to study pharmacological inhibition. Though this presents a useful approach, its limitation due to the difference between inhibition of the protein function and downregulation of the protein itself is evident. The more recent advent of the CRISPR-Cas9 RNA-guided DNA endonuclease system allows to rapidly and precisely generate loss-of-function (LOF) and gain-of-function (GOF) mutations in tumor suppressor genes, oncogenes, and other modulators of cellular transformation or drug response. In order to analyse the phenotype that results from these target manipulations, suitable assays have to be performed. These assays usually

Table 4.2 Different methods to increase or decrease the quantity of a protein target

Method	Advantages	Limitations
Ectopic overexpression	Easy Allows stable expression	Variable transfection efficiency Unphysiologic expression
siRNA knockdown	Easy Quick	Transient gene silencing Variable specificity
shRNA knockdown	Allows long term silencing Allows inducible silencing	Variable transfection efficiency Variable specificity
Anti-sense DNA	Inexpensive Quick	Variable specificity Variable efficiency
Dominant negative mutant	Allows stable suppression Allows inducible suppression	Variable transfection efficiency Unphysiologic
Constitutive active mutant	Allows stable activation Allows inducible activation	Variable transfection efficiency Unphysiologic
CRISPR/Cas9 knockout	Complete gene silencing Specific	Labour intense Lethal mutants
CRISPR/Cas9 knockin	Precise genome editing	Labour intense Variable efficiency
Small molecule inhibitor	Easy delivery	Variable specificity Variable efficiency

monitor proliferation rate, apoptosis, immortalization, anoikis, two-dimensional colony formation, colony formation in soft agar, cell migration and reliance upon growth factors. Although these cell-based assays are a cornerstone of early stage target validation, their capacity to recapitulate the complexity of human tumors is clearly limited. The artificial conditions of cell culture including the two-dimensionality, levels of oxygen, growth factors and sugar, lack of interaction with cells from the tumor microenvironment and the lack of vascularization promote adaptive processes within the cultured cells reducing their resemblance to cells of a real tumor in a human body. Great effort has been made to develop cell culture systems that represent a more realistic scenario (Gillet et al. 2013). More clinically relevant cell-based tumor models include three-dimensional cell cultures and multicellular co-culture systems capable of mimicking the in vivo microenvironment of cultured tumor cells ex vivo (Fig. 4.3).

Animal Models

Cancer involves interactions between multiple cell types, different tissues and organs. Animal models, which are whole, living systems are not only capable of modelling cell and molecular biology but also physiology. The animal models used in drug discovery and development overlap with basic research that uses these models to understand the mechanisms of the disease. Experimental approaches to target validation using whole animals overcome many of the limitations associated with cultured human cells (Smith 2003). Nevertheless, they represent non-human models and the extrapolation of data obtained by using them to human cancer is not always trivial. While mammalian animal models, predominantly mice have long been used in drug

4.2 Target Validation

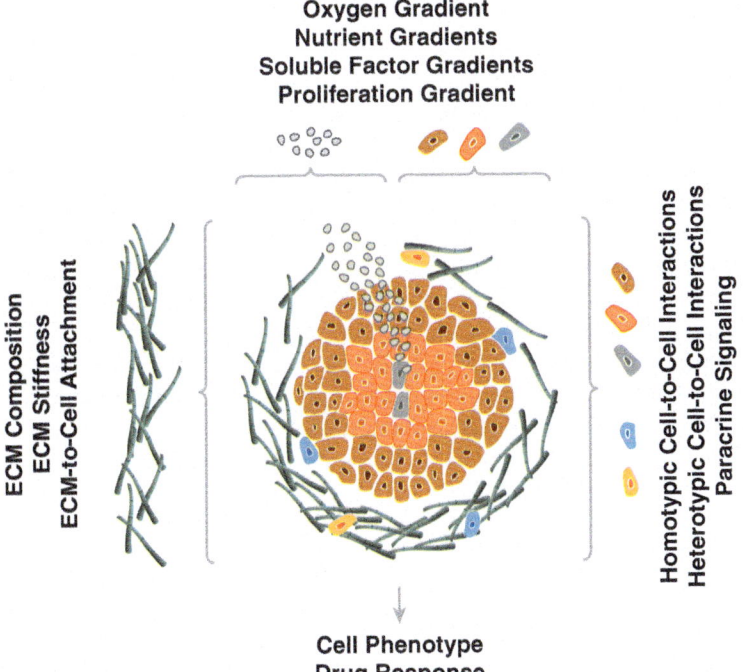

Fig. 4.3 Modern cell culture techniques try to simulate the physiological multicellular interactions and the complex with the microenvironment. Cancer cells (orange) interact with extracellular matrix proteins and glycoproteins (gray), support cells that mediate cell-cell interactions (blue), immune cells (yellow), and soluble factors (white spheres). Cell physiology is also determined by the presence of gradients of oxygen, nutrients and growth factors. Adapted from Front Pharmacol 9, 6

development, more recently researchers adopted several invertebrate and lower vertebrate animal models including worms, flies and fish. This effort was driven by the acknowledgement that these latter models are inexpensive yet powerful genetic systems which are relatively easy to manipulate and that the most important signalling pathways in human cancer are highly conserved during evolution. Invertebrate and lower vertebrate models turned out to be very useful to characterize target function in disease-relevant signalling pathways. However, most target validation in vivo is carried out in mouse models. A broad variety of different approaches have been used to faithfully model human cancer in mice. Most relevant for target validation purposes are models in which the selected target was genetically manipulated to be more abundant, altered or absent. Importantly, manipulation of the target should affect the pathology but without severe toxicity for the organism. In order to model a real world therapeutic scenario, the disease has to be established before the target is manipulated. Therefore, target validation in vivo requires tumor-bearing mice in which target function can be abrogated by genetic or pharmacological means including conditional target gene knockout or knockin, anti-sense oligonucleotides, RNAi, aptamers, antibodies or small molecule inhibitors (Fig. 4.4). An appropriate tumor model must be compatible with the mode of target manipulation. A broad variety of murine tumor models are available. They can be classified into models with spontaneous or induced tumors. Some mouse strains are known to be pre-disposed to develop certain cancers and can be used for target validation purposes. However, most in vivo target validation studies use mice in which tumors were induced either by viral infection, chemical carcinogens, tumor transplantation, transgenic expression of tumor-promoting genes, targeted gene knockout/in or somatic engineering.

Knockout mice represent the predominantly used animal model to test if the absence of the gene coding for the selected target is compatible with the physiology of the organism and capable of alleviating the cancer associated phenotypes. However, genes might play a very different role during embryonal development and adult life of an organism. The issue that the gene of interest is absent during early development of the organism in a constitutive knockout mouse has been addressed by the development of conditional gene targeting technology. The use of conditional knockout alleles

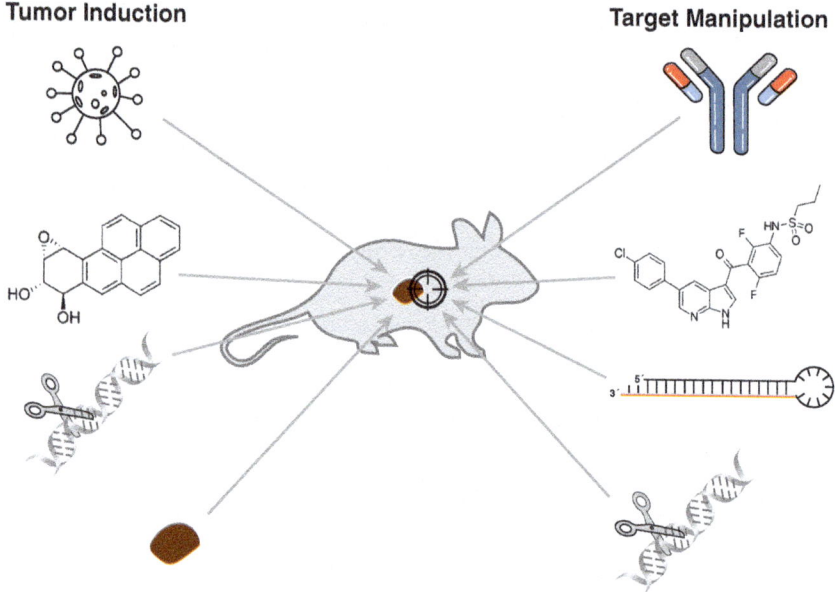

Fig. 4.4 In vivo models for target validation. The idea here is to manipulate the target of interest in a mouse and evaluate the effect of tumor growth or survival. There are different methods available to induce tumors in mice (Tumor induction, left panel). Some mouse strains spontaneously develop tumors but most of the models used in target validation rely on the artificial induction of tumors by viral infections, carcinogens, genetic manipulation of oncogenes or tumor suppressor genes or tumor transplantation. Target function (right panel) can be manipulated by pharmacological or genetic means including antibodies or small molecule inhibitors, anti-sense oligonucleotides, RNAi, aptamers or conditional target gene knockout or knockin

together with inducible Cre recombinases enables the induced genetic deletion of a target gene in the adult organism thereby more faithfully modelling therapeutic target inhibition. However pharmacological inhibition of a target does not always result in the loss of all the different functions a protein might have. In fact, the latter is the case when the selected target is absent due to gene targeting. To address this issue more sophisticated conditional knockin mouse models have been developed which carry a point mutation that results in the inactivation of a specific domain of the encoded protein. Often a chemical inhibitor does not completely shut down the function of a protein, but rather partially inhibits its activity, a scenario that might be more accurately mimicked by using shRNA-mediated gene silencing. Usually, shRNA reduces the expression of a protein, but does not eliminate it.

Human Patient Samples

The use of humans in target validation is restricted to clinical trials and the analysis of human biological specimen. These patient samples are used for translational purposes in cancer research, in particular for testing hypotheses in experimental studies. Recent developments in biotechnology open unprecedented opportunities to assess individual human genome, its expression, the complex networks of interactions between biomolecules and the functional consequences of their alterations. Therefore, scientific studies based on human specimens are becoming critical in the process of discovering new mechanisms involved in cancer, cancer progression, resistance/response to treatment and clinical outcomes. **Biobanks** are strategically important tools in several rapidly expanding domains of biomedical sciences (Hewitt 2011). The use of human samples allows for the correlation of target status including mutations, epigenetic modifications, copy number, expression level, post-translational modifications with many relevant clinical parameters including disease stage, survival and response to treatment. Normal tissue as well as blood samples from the same patient, known as matched samples can be very useful though they are not always available. Unmatched normal samples might be used instead. According to the hypothesis to be tested, different experimental approaches and sample requirements are necessary. Solid tissues are collected by biopsy or during surgical procedures. Routine clinical blood samples can be collected concurrently. Sample processing includes cryopreservation of fresh tissue or formalin or alcohol fixation and paraffin embedding. Normal clinical procedures are mainly based on fixed paraffin blocks which are useful for in situ target detection by using immunohistochemistry. However, quantification of target transcript and protein expression are preferably performed in fresh frozen tissue samples. Blood derived specimens can include plasma, white blood cells and serum. Sometimes even urine or saliva samples might be useful. Correlations of target data with clinical parameters obtained from these samples should be validated using independent cohorts.

Predictive Value of Cancer Models

The attrition rate (Box 3) of drug candidates in clinical trial clearly reflects the limitations of current experimental models used to recapitulate the features of human tumors (Sams-Dodd 2005).

Box 3
Drug attrition
Drug attrition refers to the failure of drug candidates to enter the next development phase. The processes involved in taking a candidate drug through the stages necessary to obtain marketing approval are associated with enormous challenges and a high risk of failure. There are no guarantees to success. Compared to other therapeutic areas with an average success rate of approximately 11%, oncology has one of the lowest success rates with about 5%. The reasons why drug candidates fail have significantly changed over time. Thirty years ago, pharmacokinetic liabilities (absorption, distribution, metabolism, and excretion) were the most frequent cause of attrition. The incorporation of assays to test pharmacokinetic properties in the early preclinical phase of the development process, the attrition of drug candidates due to adverse pharmacokinetic and bioavailability results decreased. Nowadays problems with efficacy and safety are the most significant reasons of drug attrition.

4.3 Lead Discovery and Optimization

A lead compound is a chemical molecule that demonstrates desired biological activity on a validated molecular target. Several approaches to identify lead compounds have been used successfully including serendipity (luck), chemical modification of active molecules, rational drug discovery and screening (Hughes et al. 2011). Even in modern drug discovery **serendipity** is still an important factor. However, serendipity does not refer to random discoveries as chance favors the prepared mind (Hargrave-Thomas et al. 2012). The classic example is when Alexander Fleming observed that bacteria were lysed in the area of mold contamination. It turned out that the mold product that inhibited the growth of bacteria was the antibiotic penicillin. A more recent example is the development of Viagra to treat erectile dysfunction. Sildenafil (Viagra) was developed to treat angina pectoris but failed in clinical trials. Nevertheless, it was observed that the drug candidate could induce marked penile erections and participants refused to return the remaining pills after the trial. Sildenafil very soon became a blockbuster drug. Modern drug discovery, however, aims to convert knowledge into treatment and reliance on serendipity is very limited. A validated target is the centerpiece of knowledge based drug discovery. However, it is not always the starting point of a drug discovery project. Often scientists start with **chemical**

4.3 Lead Discovery and Optimization

modifications of an interesting bioactive compound, an approach known as ligand-based drug design. A chemical compound might call the attention because it induces a phenotype useful to inhibit features associated with tumorigenesis such as angiogenesis, cell migration or invasion or subcellular localization of specific proteins. Medicinal chemists can introduce minor modification reducing toxicity or improving biological activity or intellectual property position and end up with a promising series of chemical compounds for clinical development. If the molecular target of these agents is unknown, target identification might involve a complex and time-consuming research effort. If the target is known, ligand-based drug design is a quite straight forward process providing a fast track to clinical development. However, the degree of pharmaceutical innovation associated with ligand-based drug design is limited, you end up with something very similar to what you start with. In oder to minimize risk associated with the drug development process, many pharmaceutical companies choose to generate so called "me too" drugs (Box 4) (Lanthier et al. 2008).

Box 4
New Molecular Entities (NMEs) and "Me too" drugs
New drugs can be grouped into "me too" drugs which display small variations of existing drugs with limited therapeutic gain and new molecular entities (NMEs) and new chemical entity (NCE). According to the U.S. Food and Drug Administration, a NCE is a drug that contains no active moiety that has been approved by the FDA before. A NME is a drug that contains an active moiety that has never been approved by the FDA. Accordingly, these categories of novel drugs are associated with a different degree of innovation from pure duplicative drugs to breakthrough drugs. Me too drugs, also called "follow-on" or copycat drugs are approved after a pioneering drug, very similar to the pre-exiting drug and not clinically superior. Me too drugs are distinct from generic drugs, which cannot enter the market until the patent on the original product expires. More innovative drugs explore novel molecular modes of action and new targets.

An increasingly important strategy in modern drug discovery is rational drug design (Mavromoustakos et al. 2011). **Rational drug design** establishes structural relationships between the biological properties and the molecular structures. The interaction of a drug with its target is based on molecular recognition. This interaction usually involves the specific attraction of chemical groups of the biological target to a small molecule drug. Rational drug design begins with the design of compounds that conform to specific requirements according to the three-dimensional structure of the target which has been resolved for many proteins. If no structural data of the target is available, the structures of homologous proteins might be known. The knowledge of the three-dimensional structure of the target allows for detailed docking studies aimed at the identification of binding mode of ligands to the protein. Docking studies are computer based analysis of drug target interactions using dedicated software packages based on genetic algorithms. Based on this technology large virtual screening campaign can be conducted against libraries that contain many million unique real compounds that are commercially available or reported. Top-ranked hit compounds from the virtual screening effort can be purchased or synthesized validated with experimental data coming from biological screenings using assays suitable to monitor the activity of the target (see below). Eventually co-crystallization of active compounds with the target can provide useful information about binding modes and to rationalize the biochemical results.

Compound Screening

Another strategy to identify drug leads is random testing of a large number of chemical agents, an approach known as **compound screening** (Hughes et al. 2011). In order to identify compounds capable of interfering with the selected target, its activity must be accurately monitored in a large-scale fashion. The process of establishing such a method is called **assay development**. An assay to measure drug-target interaction or target activity requires a readout, a measurable parameter such as light or temperature. Biological assays can be classified into biochemical assays and cell-based assays. The complexity of an assay correlates with the amount of information that can be obtained but limits the number of compound that can be tested per time (Fig. 4.5).

A **biochemical assay** is a cell-free in vitro procedure to measure the binding or activity of a target molecule. In contrast, **cell-based assays** use cells to monitor phenotypic changes such as transcriptional changes, protein expression, protein localization, protein interactions, cell viability, proliferation and migration upon the

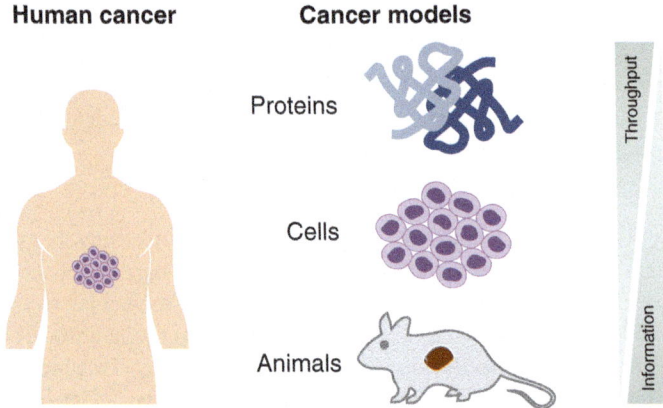

Fig. 4.5 Screening refers to an activity of random testing a huge amount of agents in an biological assay. Not all biological assays are suitable for screening as their throughput might be limited by the complexity of the biological model. Usually, whole vertebrate organisms are not used to perform screening campaigns as the possibility to test many different agents is limited. However the information one can retrieve from a model correlates with its complexity. Screening campaigns are usually performed in cell-based systems or on isolated protein targets in biochemical assays

4.3 Lead Discovery and Optimization

treatment with perturbing agents. Assays can have different designs such as end-point assays or kinetic assays, in which a single measurement after a fixed incubation time or multiple measurements over a fixed period of time are performed, respectively. Assays which determine a single parameter are called single point assays whereas multiplexed assays. Many targets have a known activity such as enzymes which convert substrates into products providing potential readouts to monitor their activities. In line with the notion that kinases play a key role in cancer cells and their inhibition is a cornerstone of modern anti-cancer therapy, kinase assays are an essential part of early drug discovery and development efforts. Accordingly, abroad spectrum of kinase assays has been developed using different readout technologies. A widely used biochemical assay to monitor the activity of kinases takes advantage of ADP accumulation upon kinase activity. In this type of assays ADP is converted by a coupling enzyme into fluorescent or luminescent chemicals that can be directly measured. In the presence of an inhibitor, the kinase generates less ADP and in turn the signal decreases. A popular and very flexible biochemical assay technology is the Amplified Luminescent Proximity Homogeneous Assay (Alpha) Screen used to measure biomolecular interactions by a bead-based proximity assay. **Alpha Screen** assays contain two bead types, donor beads and acceptor beads (Eglen et al. 2008). Binding of molecules captured on the beads leads to an energy transfer from one bead to the other, producing a luminescent/fluorescent signal. The Alpha Screen technology allows for the detection of protein–small molecule, protein–protein, protein–DNA and protein–RNA interactions. A very powerful cell based technology is **high content screening (HCS)** (Zanella et al. 2010). HCS is based on automated microscopy and multi-parameter image analysis (Fig. 4.6). Most applications use fluorescent readouts to obtain quantitative data from cell populations at a single cell level. The ability to conduct multiple independent measurements on a single selected cell-of-interest is one of the most conspicuous features of HCS. Indeed, the level of complexity comprising both changes in cellular morphology and macromolecular subcellular localization and expression level renders HCS outstanding among current approaches used for drug screening.

Usually, assays are established in bench-scale format and then down-scaled to a more economic multi-well format. Different formats of multi-well assay plates are available including 96-well, 384-well and 1536-well formats (Fig. 4.7).

Regardless of the assay design or format, before proceeding to test compound libraries, the robustness of the assay must be validated. An assay should always include positive and negative controls. DMSO is a solvent in which chemical compounds are usually dissolved and is widely used as a negative control. As DMSO is toxic to cells it should not be used beyond a concentration of 1% of the medium in cell based assays. Any treatment known to produce a signal in the assay can be used as a positive control. The most widely used parameter to assess the quality of the assay is the Z' factor (Zhang et al. 1999). The Z' factor describes the total separation between negative and positive controls minus the error associated with each type of control. The Z factor is calculated using this formula:

$$Z' = 1 - (3SD_{high\ control} + 3SD_{low\ control})/|Mean_{high\ control} - Mean_{low\ control}|$$

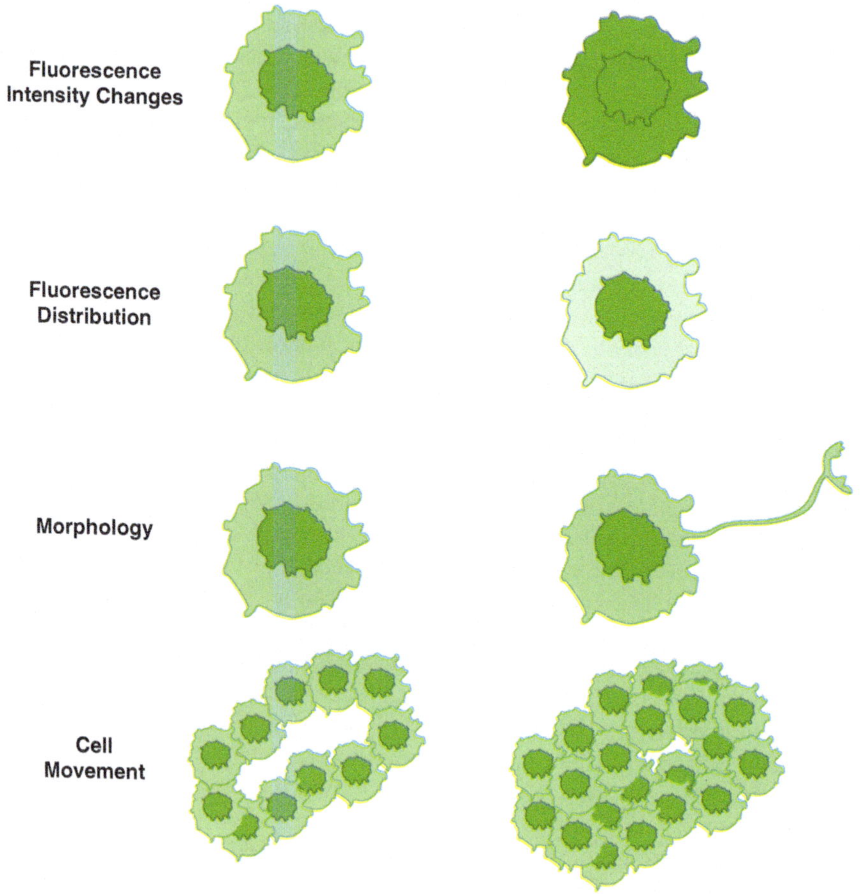

Fig. 4.6 High content screening is the combination of robotic fluorescent microscopy and automated image analysis and observes the reaction of a cell to a perturbation such as compound treatment using many different parameters including fluorescent intensity, fluorescent distribution, morphology or cell movement. For example if a protein of interest has been fluorescently labelled one can follow its intracellular localization after drug treatment

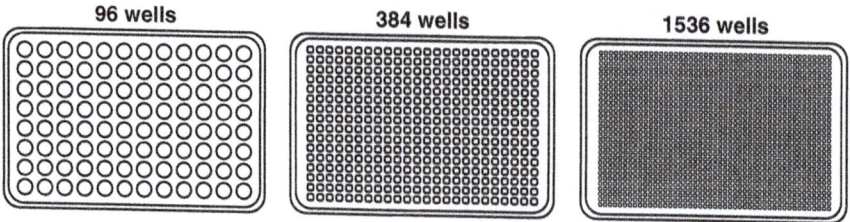

Fig. 4.7 Different formats of multiwell plates. Presented are plates with 96, 384 and 1536 wells

where σ_p = standard deviation of the positive control; σ_n = standard deviation of the negative control; μ_p = mean of the positive control; μ_n = mean of the negative control. A theoretically perfect assay would yield a Z' factor = 1, and an assay for screening is considered acceptable when its Z' factor is higher than 0.5. A small pilot screen using a small training set of compounds is useful to verify that the assay is performing reliably.

Compound Libraries
Once the assay passes all validation steps, large compound collections can be screened. Several types of compound libraries can be used, typically containing either natural products, synthetic compounds or small chemical fragments. **Fragment-based drug discovery** is based on identifying small chemical fragments with a typical size between 150 and 250 Da (Hajduk and Greer 2007). Because of their limited size and complexity, fragments are expected to bind weakly to the biological target, but medicinal chemists can easily modify them increasing size, complexity and in turn target affinity. These modifications can be rationally designed according to the structure of the target. Therefore, fragment-based drug discovery is very flexible but provides limited chemical diversity. On the other end of the spectrum are the **natural products** offering chemical diversity unmachted by any synthetic chemical collection. However, many natural products are very complex, difficult to synthesize or modify and limit the flexibility of medicinal chemists during the optimization process. Most widely used compound collections contain **synthetic compounds**. Synthetic compound refers to a substance that is made by a chemist not by nature. Many different collections of chemical compounds, called compound libraries drug screening are commercially available or owned by pharmaceutical companies. If the protein to be targeted is for example a kinase involved in a cancer signaling pathway, then rather than screening a complex library of diverse compounds, a focused chemical library would be constructed to target the ATP binding sites on the kinase enzyme. Another strategy is the systematic testing of already approved drugs to identify novel indications for existing drugs, a procedure called **drug repurposing** (Box 5) (Pushpakom et al. 2018).

Box 5
Drug repurposing
Drug repurposing also called drug repositioning refers to the application of an existing drug to a new disease outside the scope of the indication for which it has been approved. Finding new uses for approved or investigational drugs is a strategy that has several advantages. As discussed in this book, the drug development process is long, expensive and associated with a high risk of failure. Drug repurposing minimizes the risk of attrition since it is based on approved drugs for which there is no need to reproduce in-human safety studies. Drug repurposing reduces the time required to develop a drug as preclinical testing, safety and formulation have been completed. Therefore, drug repurposing

saves time and money. Assessment of efficacy for the new indication has to be performed and therefore the cost for clinical phase 3 trials which determine if the drug works in patients will be the same as for a new drug. Although nowadays drug repurposing is often performed systematically using libraries of approved or investigational drugs, or drugs that failed for reasons other than safety, serendipity has played a key role in the successful reposition of many drugs. The economically most successful repurpaosing is a good example for that. Pfizer developed Sildenafil as an antihypertensive drug, whose side effect was recognized and commercially exploited. Sildenafil was marketed as Viagra for the treatment of erectile dysfunction.

Hit Identification and Lead generation

Robotics and computational power allow researchers to test hundreds of thousands of compounds against the target. This approach has been coined **High Throughput Screening** (HTS) (Hughes et al. 2011). The primary screening is performed at a single concentration or several different concentrations, as a single point or in duplicate or triplicate. A chemical compound that produces a positive signal in an assay is called a **hit compound**. Usually, after a primary screening campaign all hit compounds are collected in a "**cherry picking**" procedure and retested using the same assay in an independent experiment. In this confirmation screen compounds chemically similar to the hit compounds but not tested in the primary screening can be included, an activity known as **hit expansion**. Confirmed hit compounds then enter the "**hit to lead**" stage a process also called lead generation. A very important step to increase the confidence in the selected hit compound is the establishment of a dose response relationship. The same assay that has been used to discover the primary hits will usually employed to generate **dose response curves** resulting in **IC50 or EC50 values** which determine the potency of a bioactive molecule (Box 6).

Box 6
IC50 and EC50 values

In pharmacology IC50 and EC50 values are used to quantify the potency of a substance. IC stands for inhibitory concentration whereas EC for effective concentration. The potency of a drug is determined in dose-response experiments in which an effect is measured upon the treatment with different concentrations of the drug. If the drug is an inhibitor the inhibitory concentration will be measured, if the drug is an activator or the readout of the assay used to analyze the effect does not measure inhibition (e.g. cell survival assays) then the effective concentration will be quantified. IC50 is defined as the half maximal inhibitory concentration, EC50 as the half maximal effective concentration or in other words a single, specified concentration required for 50% inhibition/effect.

4.3 Lead Discovery and Optimization

In order to further increase confidence in the identified hit compounds a secondary screening is performed. A **secondary screening** usually uses a different assay technology and might add complexity and in turn information. If kinase inhibitors have been identified in a cell free biochemical assay, a cell-based secondary assay might monitor the phosphorylation of a specific downstream substrate of this kinase upon treatment with these compounds. At least one cell-based assay should be employed in the hit validation process to confirm that the compound is capable of passing through biological membranes to reach the target. Once the hit compounds passed all these validation procedures, they are clustered, ranked and **prioritized**. Medicinal chemists will analyze the structure of the hit compounds and cluster them in different chemical series if several similar compounds were identified. They also identify structural elements that are known to cause problems with toxicity, solubility, absorption or synthesis. Compounds containing these notorious trouble makers should be excluded. Compounds with too many rings, multiple chiral centers as well as inorganic and heteroatom compounds will also be eliminated too. In more recent compound libraries frequent hitter, false positive and promiscuous compounds should not be present. The difference between a good ligand and a successful drug is that the latter is not only potent against the intended target (as a good ligand), but also exhibits good physical and chemical properties. Many parameters have to be taken into consideration when ranking the remaining hit compounds (Table 4.3).

First and foremost, potency and selectivity of the hit compounds are analyzed. **Potency** refers to the concentration of the compound required to achieve a defined biological effect and is quantified in the IC50 or EC50 value. Hit compounds should have a significant potency, at least in the lower μM range, otherwise their further development might not be worth the effort. **Selectivity** refers to the effect of a compound on a desired target versus an undesired target. If the drug discovery project is aimed at the identification of an inhibitor of a specific kinase, the hit compound should reduce the enzymatic activity of the target without affecting the other 500 kinases encoded by the human genome. Highly promiscuous compounds should be identified as soon as possible as lack of selectivity represents one of the liabilities that are most difficult to fix later on during the development process. Commercial panels of hundreds of targets are available to evaluate selectivity and discard promiscuous compounds. Another important parameter to evaluate the quality of the hit compound and the potential of its development is tractability. **Tractability** refers to the feasibility of synthesis and flexibility towards chemical modification. For a phar-

Table 4.3 Parameters to be considered to select compounds for their further development

Parameter	Definition
Potency	Concentration of the compound required to achieve a defined biological effect
Selectivity	Effect of a compound on a desired target versus an undesired target
Drug-likeness	Properties which determine similarity to known drugs
Tractability	Feasibility of synthesis and flexibility towards chemical modification
Patentability	Possibility to obtain intellectual property for a drug candidate

maceutical company the **Patentability** is a very important aspect to decide whether an investment in the development of a hit compound could produce profit. The evaluation of the intellectual property position will be performed by using specialized databases. **Drug-likeness** has emerged as a very useful parameter to evaluate the potential of a hit compound to become a successful drug. The assessment of the drug-likeness of a hit compound is performed based on computational analysis of its physicochemical properties. The **Lipinski's Rule of Five** emerged as an important guideline for drug-likeness. These parameters of drug-likeness are widely used as filter in the hit prioritization process and were published by Cristopher Lipinsky in 1997 (Lipinski et al. 2001). Lipinsky analysed the physicochemical properties of clinically tested drug molecules and showed that the great majority of them follow certain rules. Lipinski's Rule of Five consist of 4 rules but all numbers that quantify the corresponding properties are multiples of five, and that is the origin of the name (Table 4.4). Lipinski found that most oral drugs used in the clinic are relatively small and moderately lipophilic and extrapolated this observation as desirable properties for drug candidates.

Intellectual Property

In order to protect a compound from identical or related competitor products a strategy for intellectual property (IP) protection is usually developed at early phases of the drug development process before a candidate drug enters clinical assessment. There are several types of intellectual property protection including patent, exclusivity, copyright or trademark. A patent provides exclusive rights to an inventor when the claims in the patent support the criteria of global novelty, non-obviousness and usefulness. IP protection enables the inventor to reap commercial benefits from the invention. Strong IP protection is essential for pharmaceutical companies to further invest money in the development of a candidate drug. Therefore, searching databases with information about patents relevant for the candidate drug will be performed routinely at the preclinical stage to address questions related to IP protection. Usually the protection of IP through a patent lasts 20 years from the date of its application. The strongest IP protection can be obtained if the patent claims cover a new chemical or a new molecular entity.

Table 4.4 Lipinski's Rule of Five, molecular weight in Daltons, log P measures how hydrophilic or hydrophobic a compound is, the hydrogen bond donor is an electronegative atom covalently bond to a hydrogen atom

Lipinski's rule	Parameter	Value
1	Molecular weight	<500
2	Octanol-water partition coefficient, log P	<5
3	Hydrogen bond donor groups	<5
4	Hydrogen bond acceptor groups	<10

The hydrogen donor pulls the covalently bonded electron pair closer to its nucleus leading to a hydrogen atom with a partial positive charge. A hydrogen acceptor is the atom or molecule that interacts with the partially positively charged hydrogen atom with its electron pair to form a hydrogen bond

Lead Optimization

The lead generation process results in a lead compound or drug lead (Hughes et al. 2011). A **lead compound** is a chemical molecule that demonstrates desired biological activity on a validated molecular target. Its chemical structure is used as a starting point for chemical modifications. Newly identified compounds may have poor drug-likeness and may require chemical modification to become drug-like enough to be tested biologically or clinically. During the lead optimization process medicinal chemists attempt to improve the physical and chemical properties of a lead compound introducing small structural modifications. These modifications are guided by the early evaluation of the **pharmacokinetic** and **pharmacodynamic** properties of the lead molecule (Fig. 4.8; Sheiner and Steimer 2000). Pharmacokinetics and pharmacodynamics study the interactions of the drug with the body. Pharmacodynamics is the effect that the drug has on the body including its therapeutic effect and associated toxicities while pharmacokinetics describes what the body does to the drug including absorption, distribution, metabolism and excretion.

The lead optimization process consists of iterative cycles of chemical design and biological assessment aimed at the selection of a drug candidate for preclinical development. In order to optimize the lead compound, it is important to understand precisely which structural features are responsible for its biological activity. Therefore, one of the first activities in the lead optimization stage is the identification of the pharmacophore. The **pharmacophore** is the precise part of the lead compound that is responsible for its biological activity. The study of the relationship between a molecular structure of a chemical compound and its biological activity is called **Structure-Activity-Relationship (SAR)** analysis (Perkins et al. 2003). In order to establish a SAR for a lead compound a quantitative biological assay is required.

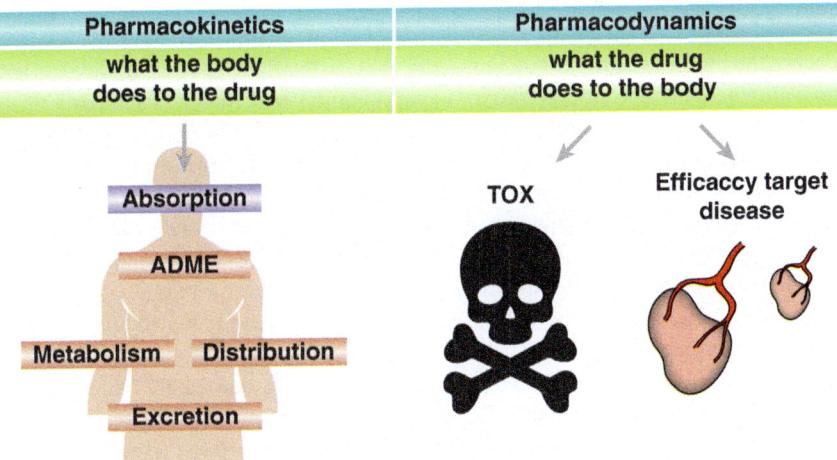

Fig. 4.8 Pharmacodynamics refers to the effect that the drug has on the body including its therapeutic effect and associated toxicities while pharmacokinetics describes what the body does to the drug including absorption, distribution, metabolism and excretion (ADME)

Based on such an assay, structurally different derivatives with small synthetic modifications can be tested and their activity quantified. Initial information about SAR might be obtained from the primary screening if analogues of the hit compound were tested or from hit expansion efforts during the confirmation screening. **Structure-Property-Relationships (SPR)** is a very similar procedure which is performed to improve the physical and chemical properties of the lead compound. SAR and SPR analysis usually involves iterative cycles of synthesis and screening to identify those chemical modifications which help to improve important properties such as solubility, metabolic stability, potency or specificity. On the other hand, SAR analysis allows for the elimination of excessive functionality which is not required for the biological function. This might be important to reduce toxicity and cost of synthesis. During the lead optimization process pharmacokinetic and pharmacodynamic properties of the lead molecule are assessed by in vitro assays and improved by chemical modifications.

Early ADME Analysis
In order to understand the importance of the physical and chemical properties of the lead compound it is useful to follow a drug through the body from administration until excretion and most importantly until it reaches the intended target. From the mouth to the target organ, the drug has a challenging journey ahead crossing several barriers, being exposed to hostile environments and attacked by enzymes. The drug has to pass through several types of membranes including cell membranes and if the target organ is the brain even the blood/brain barrier which provides a protective environment for this organ. Size and lipid solubility are important factors for cell penetration and even more for passing through the blood/brain barrier. Usually the walls of capillaries do not represent a major obstacle for drugs as pores between the cells are larger than most of them. When a drug is swallowed and reaches the stomach it is exposed to a very acid pH of 2. Then the pH gradually increases in the intestine and when the drug reaches the blood stream it encounters a pH of 7.4. The blood stream takes the drug to the liver where metabolic enzymes try to get rid of the drug molecules. Finally, in the target organ, a sufficient amount of drug has to reach the tumor to destroy the cancer cells. This is an important consideration. If the potency of a drug has been determined with the half maximal concentration of 100 nM in in vitro assays (IC50), then the drug has to accumulate at this concentration for the period of treatment within the tumor until the next dose of the drug is administered. So even the most potent and specific inhibitor would be clinically useless if it was not bioavailable or stable as it never would reach the tumor. In order for a lead compound to survive this journey, its **pharmacokinetic** properties have to be optimized. Importantly, a successful drug must be absorbed into the bloodstream, distributed to the proper site of action in the body, metabolized efficiently and effectively and successfully excreted from the body. These pharmacokinetic or **ADME (Absorption, Distribution, Metabolism and Excretion)** properties describe the disposition of a compound within an organism and influence the activity of the compound as a drug (Fig. 4.9; Hodgson 2001).

Remarkably, the pharmaceutical industry managed to reduce drug attrition rate due to pharmacokinetic reasons 20 years ago by introducing relatively simple in

4.3 Lead Discovery and Optimization

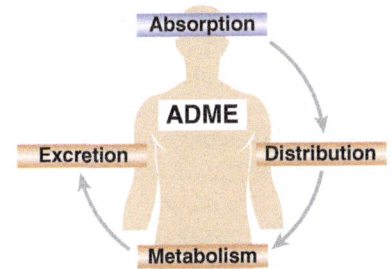

Fig. 4.9 ADME stands for Absorption, Distribution, Metabolism and Excretion. ADME studies are critical to develop new medicines. Essential aspects of these studies include bioavailability and clearance of the drug, metabolism and drug–drug interactions

vitro assays to guide medicinal chemistry during lead optimization. In modern drug discovery ADME properties of lead compounds are determined in early phases of the process. Nevertheless, the analysis of pharmacokinetic Early ADME assays assess the solubility, lipophilicity, membrane permeability and metabolic stability of the lead compound as well as its capacity to bind plasma proteins and inhibit or induce enzymes that are essential for the metabolism of many drugs (indicative of possible drug-drug interactions).

Absorption

One of the first tests is the screening for oral-absorption potential. Absorption is the process by which a drug moves from the site of administration into the blood. There are several **routes of administration** for a drug including the **parental** routes, by topical or respiratory administration or by injection and the **enteral** routes by rectal or oral administration (Fig. 4.10).

Oral delivery is the most preferred and convenient route of drug administration. Small molecule compounds can cross lipid bilayers by passive or active transport.

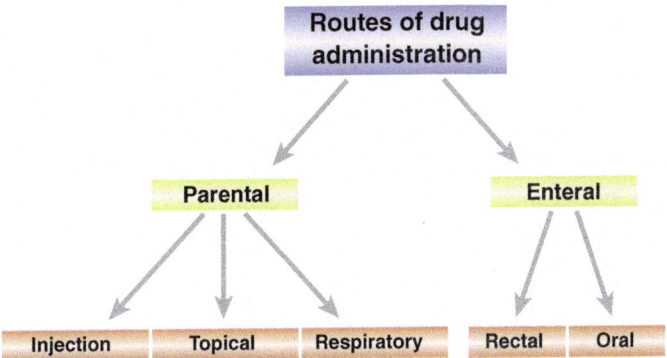

Fig. 4.10 There are several routes of administration for a drug. These routes can be classified into enteral or parenteral. Enteral routes include oral and rectal administration and deliver the drug system-wide through the gastrointestinal tract whereas parenteral administration delivers the drug by routes other than the gastrointestinal tract. Parenteral routes include topical drug administration and administration through injections or the respiratory system

Drug absorption usually occurs by passive diffusion across cell membranes. Membranes consist of a lipid bilayer with embedded proteins. The core of the bilayer represents a lipophilic environment and, as a consequence, lipid soluble compounds diffuse through membranes at a higher degree than hydrophilic compounds. As passive diffusion represents the major absorption pathway permeability assays measuring passive diffusion are considered as surrogate for oral absorption. Two permeability assays have become prevalent in recent years the **parallel artificial membrane permeability assay (PAMPA)** and the Cell based Caco-2 permeability assay (Faller 2008). PAMPA is a test to measure passive permeability in the absence of transporters or efflux systems. PAMPA was first described in 1998 and is based on the diffusion of the test compound from a donor buffer through an artificial membrane into an acceptor buffer. The drug concentrations before and after diffusion are determined and the effective permeability calculated. The capability to quantify a compound of interest in buffers and body fluids is of key importance to pharmacokinetic studies. Liquid chromatography/Mass spectrometry (LC/MS) (Box 7) is the main workhorse to analyse the ADME properties of a lead compound, nowadays (Kostiainen et al. 2003).

> **Box 7**
> **Liquid chromatography/Mass spectrometry**
> In order to characterize the ADME properties of a candidate drug, the capacity to identify and quantify a specific chemical substance in cells, cell lysates, tissues and body fluids is critical and can be accurately achieved by using mass spectrometry, which is an analytical method that measures the masses within a sample. Mass spectrometry is the workhorse of pharmacokinetic analyses. Mass spectrometry is often combined with the physical separation capabilities of liquid chromatography (or HPLC). Liquid chromatography/Mass spectrometry (LC/MS) is a tandem technique in which liquid chromatography separates mixtures of different components and mass spectrometry identifies the structural identity of the individual components. LC is based on differences in partitioning behavior between a mobile phase which carries the sample and a stationary phase within a column. Afgter the physical separation MS ionizes the chemical compounds to generate charged molecules and measure their mass-to-charge ratios. A MS instrument consists of three modules, an ion source, a mass analyzer and a detector. The atoms or molecules in the sample can be identified by correlating known masses to the identified masses.

The cell-based **Caco-2 permeability assay** measures the sum of passive and active permeabilities. Caco-2 is a human colon colorectal adenocarcinoma cell line. These epithelial cells express transporter and efflux proteins as well as metabolic enzymes and form very tight junctions between cells when grown as a monolayer. These properties make them an ideal system to model paracellular movement of compounds across the monolayer recapitulating many aspects of the human intestinal epithelium. In this assay, the test compound is added to one of two different chambers filled with buffer. As the two chambers are separated by a Caco-2 monolayer grown on a semipermeable plastic support the concentration of compound can be quantified by LC/MS in both chambers after different incubation periods and the rate of permeability expressed as the apparent permeability coefficient calculated.

Distribution

Distribution refers to the movement of a drug from the blood to the tissues and from the tissues to the blood. The movement of a drug through the body is an important parameter that influences how fast and at which concentration it is present at its sites of action. The distribution of a chemical substance mainly depends on four factors, namely the blood flow, the tissue solubility of the drug, binding of the drug to macromolecules and the ability to cross special barriers. Tissues that receive a low degree of blood flow accumulate the drug slowly whereas tissues with a high degree of blood flow have a faster rate of uptake. At the low end of the spectrum is the adipose tissue while the liver, kidneys and brain are the organs most supplied by blood. On the other hand, the brain and the adipose tissue have a high lipid content and dissolve a higher concentration of lipophilic agents delaying the movement of a drug to its sites of action. If a compound is very sticky and strongly binds to plasma proteins abundant in the blood, it diffuses less efficiently and its presence at tumor sites is limited. Many drugs bind to **plasma proteins** in the blood stream (Smith et al. 2010). Usually, basic drugs bind to the α-1 acid glycoprotein or lipoproteins while acidic and neutral drugs primarily bind to albumin, the most abundant protein in the blood plasma of humans. Only the free drug can exert the pharmacological effect. Therefore, plasma protein binding decreases distribution. Finally, the distribution of a drug might be limited by special barriers like the blood/brain barrier or the blood/testis barrier. Distribution is measured as a volume of distribution (V_D). V_D measures the fluid volume that would be required to contain the amount of drug present in the body at the same concentration as that measured in the plasma. V_D can be calculated by using the following formula:

$$V_D = \text{total amount of drug in the body/concentration of the drug in blood plasma}$$

Compounds that extensively bind to plasma proteins have a low volume of distribution. A widely utilized method to measure plasma protein binding of a lead compound is **equilibrium dialysis** which is still considered as the gold-standard of protein binding determination. In this assay, the test compound is added together with plasma in the sample chamber that is separated from the assay chamber containing buffer by a dialysis membrane which retains the proteins. The test compound is small enough to pass freely through the membrane. The concentrations of unbound and total drug in plasma can be estimated after the establishment of an equilibrium state between the two chambers after a period of incubation and a defined temperature. Drugs with known values of plasma protein binding such as Propranolol, Warfarin or Diclofenac can be used as internal controls. An alternative and faster method to measure the binding of a chemical substance to plasma proteins is **ultrafiltration**.

The test compound is incubated in plasma for one hour to allow drug binding and then centrifuged to separate the supernatant through a semipermeable membrane for the quantification of unbound compound. Only a small amount of compound is required for ultrafiltration assays, but the method is known to be susceptible to unspecific protein binding.

Metabolism

Metabolism refers to the transformation of drugs to make them more hydrophilic allowing them to be excreted more easily by the kidneys. The body regards drugs as foreign substances, not produced naturally and wants to get rid of them. Urine is a major pathway of excretion eliminating water soluble substances. Oral drugs however use to have a limited solubility in water. As oral drugs have to be capable of crossing membranes, they have to be a little bit lipophilic. In fact, the selection of drug leads is a balancing act between solubility and lipophilicity. Therefore, the body has to chemically transform these drugs into more polar structures and as a consequence more water-soluble to excrete them. The biotransformation is executed in two different phases, also called phase-I-metabolism and phase-II-metabolism. The drug molecules are first modified by specific phase-I enzymes which mediate oxidation, reduction, hydrolysis or the. removal or addition of an active group. Then the phase-II enzymes couple polar groups such as sulphates and glucuronide which facilitate excretion to existing (or phase I formed) conjugation sites. These processes mainly take place in the liver, the main organ of detoxification (Fig. 4.11). Some drugs such as morphine, propranolol or diazepam are significantly metabolized in the liver before they reach the rest of the body. This effect is known as hepatic first pass.

The cytochrome P450s (CYP450) are the most important enzymes in detoxification of drugs (Nebert and Russell 2002). Almost every drug is processed by some of these enzymes. The human genome codes for 57 cytochrome P450s genes, but five isoenzymes, 1A2, 2B6, 2C9, 2D6, 3A4 metabolize about 90% of all drugs (Fig. 4.12).

CYP450 proteins containing a heme cofactor whose iron is used to oxidize molecules are responsible for most phase I reactions. The consequence of metabolism is not always drug inactivation. The resulting metabolites may have equal activity to the drug, no or reduced activity, increased activity, which is the case of prodrugs or toxic properties, not seen with the parent drug. The main issues associated with CYP450 in drug development are metabolic stability and CYP inhibition/induction. Drug researchers try to avoid excessive metabolism of a drug candidate as it destabilizes the molecule. A drug which is rapidly metabolized, may have no therapeutic benefit because it never reaches the tumor. If a drug inhibits a specific CYP450 it may delay the metabolism of another drug molecule which in turn accumulates leading to severe toxicity. This effect is known as drug-drug interaction and is considered as a major cause of adverse drug effects. Interestingly, grapefruit juice contains a compound that inhibits the activity of CYP3A4, an enzyme that metabolizes many commonly used drugs. Therefore, you should avoid drinking grapefruit juice when taking these drugs. On the other hand, if a drug candidate induces the expression or activity of a specific CYP450 causing the rapid metabolism of another drug which as

4.3 Lead Discovery and Optimization

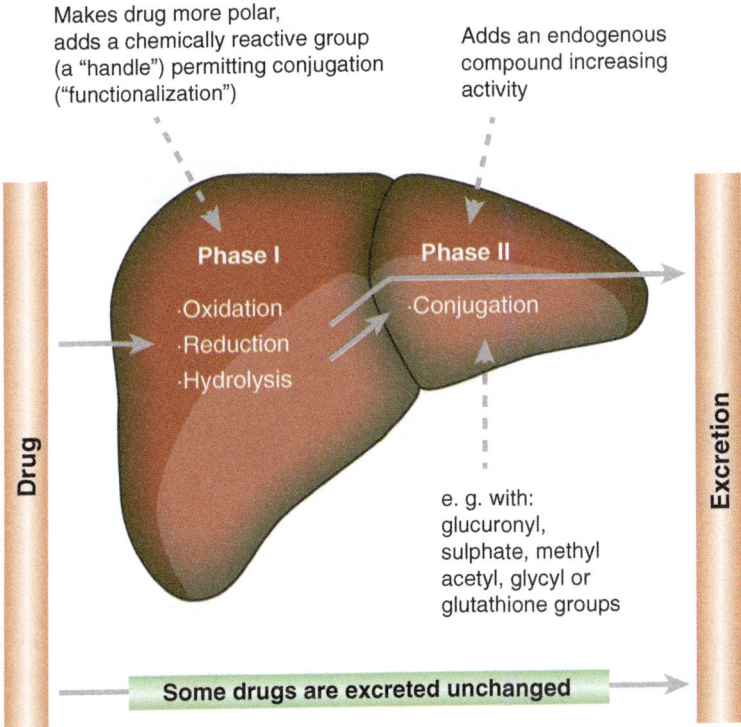

Fig. 4.11 Metabolism is the conversion of a substance into another form by metabolic enzymes in the body. The Liver is the primary organ of drug metabolism. Metabolism of the majority of drugs occurs in two phases. During phase I metabolism, drugs will be modified by adding chemically reactive groups to make them more polar and allowing for conjugation. Phase 2 metabolism is carried out by specific enzymes mainly transferases that add chemical groups such as glucuronyl, sulphate, acetyl and glutathione groups

a consequence loses its therapeutic benefit, the outcome might be extremely severe. Therefore, any liabilities with CYP450 should be identified in the early phases of dug development. The implementations of in vitro assays facilitated the measurement of metabolic stability, CYP inhibition as well as CYP induction and these tests are standard procedures in lead optimization (Walsky and Obach 2004). Metabolic stability of a lead compound is determined in liver extracts, CYP inhibition in cell extracts from CYP overexpressing cells and CYP induction in intact liver cells. In vitro assays to monitor the metabolic stability of a lead compound are very simple and straight forward. They are usually based on subcellular fractions of liver extracts such as S9 for microsome fraction from mice, rats and most importantly humans. These fractions are prepared by using centrifugation procedures and contain different metabolic enzymes. However, CYP450 are known to be highly polymorphic leading to many individual differences in CYP450 expression and activity. In order to address this issue, big pools of post-mortem human liver extracts from different sexes, ages

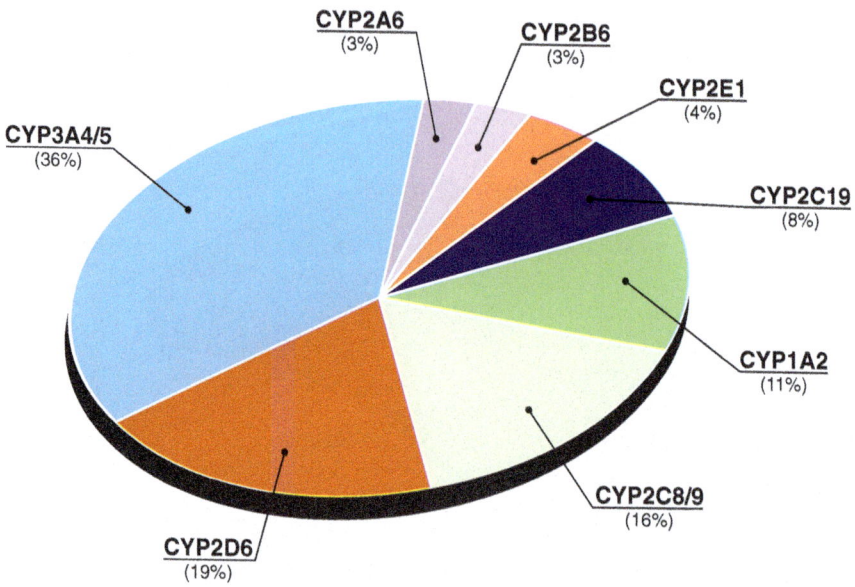

Fig. 4.12 Cytochromes P450 (CYPs) are enzymes that contain heme as a cofactor and oxidize their substrates. Among the many different isoforms of CYPs, CYP3A4/5, CYP2D6, CYP2C8/9, CYP1A2 and CYP2C19 metabolize about 90% of drugs. Percentage of drugs metabolized by a specific CYPs are provided

and ethnic groups are used. To test the metabolic stability of a lead compound in vitro, liver microsomes are incubated together with the test candidate for different periods of time. After 15 and 30 min of incubation at 37 °C the concentration of the remaining test compound is measured by using LC/MS and expressed as percentage of the initial drug concentration. Rapidly metabolized lead compounds are either discarded or handed over to medicinal chemists for their modification to stabilize them. Most assays to monitor **CYP450 inhibition** use cDNA-expressed human CYP450 isoenzymes as the enzyme source. They are based of the conversion of an artificial substrate that is converted by a specific CYP450 isoform into a product that emits a detectable light signal. The strength of this signal depends on the activity of the CYP450 enzyme and decreases in the presence of a lead compound that inhibits its enzymatic activity. CYP450 induction testing in based on the use of immortalized hepatic cell lines such or primary cultures of human hepatocytes. To compensate for the above mentioned individual differences in CYP450 enzymes, hepatocytes from at least three individual donor livers should be used for these experiments. Cells are exposed to different concentrations of the lead compound or positive control inducers usually for two or three days. Then transcript or protein expression of the most relevant CYP450 isoforms is determined by using immunoblotting or RT-PCR, respectively. Alternatively, assays to measure the enzyme activity can be performed.

4.3 Lead Discovery and Optimization

Metabolism is one of two ways by which the body can remove the parent drug. The other is excretion.

Excretion

Excretion in pharmacokinetics refers to the removal of drugs from the body. Drugs can be eliminated from the body by several routes such as urine, bile, sweat, tears, saliva, milk, and stool. By far, the most important excretory pathway involves the kidneys. The kidneys filter blood and produce urine. One kidney contains approximately one million nephrons, which are the functional units of the kidney. The blood flows at very high pressure into a small knot of blood vessels called the glomerulus which is a nonselective filter permeable to all the compounds whose molecular weight is below 65,000 Dalton. Accordingly, the molecular weight of small molecule drugs (MW < 900) is not a limiting factor and therefore, the concentration of drugs in the filtrate is identical to the unbound drug in the plasma. The filtrate is collected in the renal tubules that transport urine into the collecting tubule, whereas the filtered blood moves on through capillaries into the renal vein that exits the kidney. However, before the filtered blood reaches the renal vain, remaining unneeded substances are actively secreted from the peritubular capillaries into the tubule of the nephron. On the other hand, before the filtrate reaches the collecting tubule useful substances are reabsorbed back into the blood stream through the Capillaries (Fig. 4.13). Tubular secretion and tubular reabsorption can alter the final concentration of the drug molecule in the urine quite significantly.

In summary, the nephron performs glomerular filtration, tubular secretion and tubular reabsorption, three processes which determine the amount of the excreted drug expressed by the following formula:

$$\boxed{\text{excreted amount} = \text{filtered amount} - \text{reabsorbed amount} + \text{secreted amount}}$$

Excretion is the net result of these three processes which define what goes into the urine. Only hydrophilic molecules can be efficiently excreted by urine. Drugs are eliminated by the kidneys through glomerular filtration and tubular secretion either as unchanged agents or as metabolites. Lipophilic drug molecules have to be biotransformed into hydrophilic drug metabolites to be excreted. Analysis of excretion is usually carried out in rodent and nonrodent animals to understand how the parent and metabolites are excreted.

Early Toxicity Assessment

It is important to identify safety liabilities of lead compounds to remove the most toxic drugs or to guide chemists in designing modified compounds with a more favourable toxicity profile before they are tested in more expensive animal model systems (Li 2004). The drug safety profile and risk tolerance depends on the therapeutic area. The often life-threatening nature of cancer led to an increased risk tolerance for drugs used in oncology. Safety issues associated with the toxicity affecting liver, the cardiovascular system, the central nervous system, or the haematological system are a major cause of failure of clinical trials and withdrawal from sale after marketing

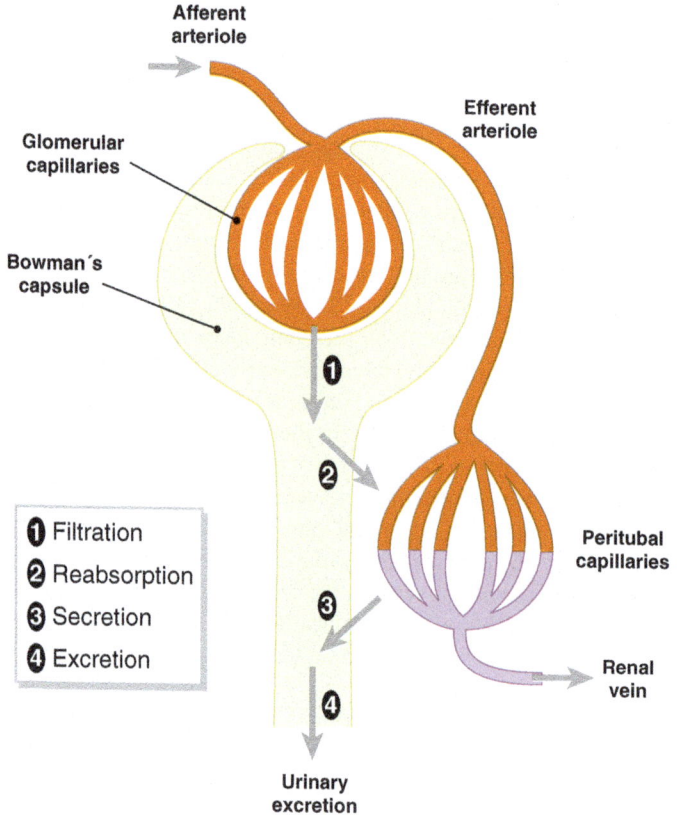

Fig. 4.13 Most drugs are excreted by glomerular filtration and by active tubular secretion that take place in the nephron. The nephron is the smallest functional unit of the kidney and is composed of the glomerulus, the Bowman's capsule and the renal tubule

drug. This is a significant problem for the pharmaceutical industry as the attrition of drug candidates due to toxicity issues has increased during the last decades. This reflects the fact that the implementation of relatively simple and reliable in vitro assays to predict toxicities in humans has not been achieved in the same way as for early ADME assessment. Nevertheless, several assays to predict safety liabilities are carried out in the lead optimization stage usually including early assessment of hepatotoxicity and cardiotoxicity.

Hepatotoxicity

The liver is the main drug metabolizing organ in the body. The liver is responsible for breaking down or modifying toxic substances including drug molecules but is also susceptible to damage cause by these agents and their metabolites. Drug induced liver injury is a major cause for drug withdrawals from the market (Lee 2003).

4.3 Lead Discovery and Optimization

Drug attrition and withdrawal due to hepatotoxicity lead to huge financial losses for pharmaceutical companies. Examples of approved drugs which have been withdrawn from the market because of their hepatotoxicity include troglitazone, nefadazone and trovafloxacin. The development of in vitro and in vivo models that recapitulate human liver physiology has turned out to be a consistent challenge. As many drug metabolites specifically generated by human metabolic enzymes are hepatotoxic, any method to model liver toxicity should include metabolically competent human hepatocytes. In vitro models rely on cultured hepatocytes which are known to lose many traits present in liver cells very fast upon cell culturing. The standard in vitro test to identify potential drug induced liver injury liabilities of lead compounds cell viability assays using established cells lines such as HepG2. HepG2 is a widely used cell line derived from a patient with well-differentiated hepatocellular carcinoma. More advanced in vitro methods use primary hepatocytes in coculture with other cell types in two or three-dimensional formats and multiparameter readouts based on high content analysis. Control agents should always include known hepatotoxic compounds.

Cardiotoxicity

Cardiotoxicity is known to be associated with many drugs used in oncology including standard chemotherapeutic drugs like the anthracyclines or targeted drugs such as Trastuzumab (Pai and Nahata 2000). Just as hepatotoxicity, cardiotoxicity is a leading cause of attrition of drug candidates and accounts for about 25% of withdrawal of approved drugs from the market in the US. Cardiotoxic drugs exert their damaging effect to the cardiovascular system by several known molecular mechanisms including damage to mitochondrial damage, disruption of kinase signalling and inhibition of cardiac ion channels. Almost thirty years ago, an increasing number of drugs were reported to cause cardiac arrhythmias. The most common drug-induced arrhythmia is torsades de pointes caused by prolonging electrical depolarization and repolarization of the ventricles. Torsades de pointes (TdP) is a form of ventricular tachycardia that might result in ventricular fibrillation and cardiac death. Drugs that induce torsades de pointes are believed to inhibit the subunit of a potassium ion channel encoded by Ether-à-go-go-Related Gene (hERG) in humans. In order to detect liabilities associated with hERG inhibition early in the drug development process, in vitro hERG assays have been implemented in the lead optimization process. Most of these assays are based on the ectopic expression of hERG channel in human embryonic kidney (HEK) or Chinese Hamster Ovary (CHO) cells in which hERG inhibition can be monitored by several methods. These methods include the measurement of hERG current by patch clamping, a displacement assay using radio-labelled channel blocker or the use of fluorescent dyes to measure the membrane potential.

4.4 Pre-clinical Drug Development

Preclinical development is the process of taking an optimized lead compound through the stages necessary to allow human testing (Hughes et al. 2011). During the preclinical phase of drug development information about safety and efficacy of a drug candidate is obtained to decide if it is safe and efficient enough for human testing. Preclinical studies include in vitro and in vivo experiments on pharmacokinetics, pharmacodynamics, and toxicology that have to comply with the guidelines dictated by **Good Laboratory Practice** (Box 8). Based on the results of these preclinical studies a decision will be taken whether a drug candidate can enter clinical studies. Before any clinical trial can begin, the sponsor, usually a pharmaceutical company must obtain permission to test the candidate drug in humans filing an **Investigational New Drug (IND)** application. The application is reviewed by regulatory authorities to make sure people participating in the clinical trials will not be exposed to unreasonable risks. Studies in humans can only begin after IND is approved. A significant effort has been made to harmonizer the requirements for human clinical evaluation amongst international regulatory authorities including the Food and Drug Administration in the U.S., the Therapeutic Products Directorate in Health Canada, and the European Medic EMEA and other worldwide bodies.

Box 8

Good Laboratory Practice (GLP)

Preclinical experiments pave the way for candidate drugs to be tested in humans. In order to prevent any health risk for study participants, a strict quality control has to ensure that the obtained pre-clinical data are reliable. GLP is a control system for non-clinical research to ensure quality, reliability, reproducibility and uniformity of safety and efficacy studies for the development of new drugs. GLP defines the responsibilities of the personnel involved in the preclinical drug development process, as well as the requirements for the facilities, equipment, reagents, materials, test procedures, and the reporting and storage of obtained data. GLP was established in the 1970s as a reaction to several cases of fraud.

During preclinical development of a drug candidate four questions have to be answered before a drug candidate can be tested in humans (Table 4.5).

Table 4.5 Questions to be addressed by pre-clinical studies

Number	Question
1	Is the drug candidate safe?
2	Does it work?
3	Can it be delivered?
4	Is its manufacturing viable?

4.4 Pre-clinical Drug Development

Preclinical Evaluation of Toxicity—Is it Safe

One of the major challenges in drug development is the accurate prediction of drug toxicity in humans. Classical toxicology is based on the concept that "all things are poison and nothing is without poison. Solely the dose determines that a thing is not a poison" proposed by Paracelsus in the 15th century. Therefore, the toxicity of an agent is determined by establishing a dose-response relationship, and safety of a drug candidate is estimate by the ratio between the dose required for the therapeutic effect. However, without a mechanistic understanding of drug toxicity, the prediction of safety in the human population based on experimental data obtained in vitro and in animals is not always accurate. The tragedy caused by the drug thalidomide in the early 1960s marked a turning point in toxicity testing of drug candidates (Kim and Scialli 2011). **Thalidomide** was sold in many countries to treat nausea in pregnant women and caused birth defects in thousands of children. The observation that Thalidomide did not cause malformation in rodent animal models and in monkeys only at a much higher dose than in humans prompted regulatory agencies to change the toxicity testing protocols leading to the requirement to test a wide range of doses, using several exposure periods and to include non-rodent species. Nowadays, the standard approach to toxicity testing includes acute, sub-chronic and chronic exposure in three animal species. Regulatory authorities generally require that drugs are tested in both a rodent and a non-rodent mammalian species. Usually, these tests are carried out in mice, rats and dogs. Regulatory agencies acknowledge that the evaluation of oncology drugs is a balancing act between risk and benefit for patients and therefore accept a higher degree of toxicity than for non-oncology drugs. Drug safety is initially evaluated in **acute toxicity tests** where a single, high dose of a substance is given to animals. Animals are observed for fourteen days and detailed records are made of body weight, behavior, temperature, respiratory and cardiovascular parameters. After two weeks animals are killed and dissected to screen for any further signs of organ toxicity. Acute studies are usually performed on a rodent and a non-rodent mammalian species early on in the development of a drug. The main objective of acute toxicity analysis in early preclinical drug development is to determine the range between the dose that causes no adverse effect and the dose that is life-threatening. These studies reveal the **severe toxicity dose (STD)** which refers to the drug dose that causes severe toxicity (or death) in rodents. The calculation of the starting dose to test a drug candidate for the first time in humans is based on the STD value. Usually, the first dose used in humans is the dose that causes severe toxicity in 10% of rodents (STD_{10}) adjusted to the human body surface area in mg/m^2. A compound which survives the acute toxicity testing will be considered for repeat-dose toxicity testing in animals. Repeat dosing studies are generally carried out in two species e.g. rats and dogs with treatment duration, schedule, route of administration and clinical formulation of the drug that mimics the intended clinical study. A range of doses will be tested including the **maximum tolerated dose (MTD)** which is the highest dose of a treatment that will produce the desired therapeutic effect without causing unacceptable side effects. Several clinical parameters will be assessed during the treatment period including behavior, body weight, blood pressure and electrocardiogram (in dogs), ophthalmoscopy, clinical pathology and histopathology. These tests also assess the reversibility of severe toxicities examined off-therapy and select target organs that need to be closely monitored in clinical trials. Drugs with toxicity only in humans and not in non-human animals should be detected in the clinical trials. Unfortunately, due to several limitations in the design of clinical trials this is not always the case. That is one of the reasons why 2.9% of the marketed drugs were withdrawn from the market during the last four decades.

Preclinical Evaluation of Efficacy—Does it Work
Efficacy evaluation of an anticancer drug candidate involves testing the impact on the viability of a broad variety of cancer cell lines, xenograft experiments in nude mice and experiments in more sophisticated genetically engineered mouse model. The predictive power of preclinical models to evaluate the efficacy of a drug candidate depends on the capacity of the experimental system to faithfully recapitulate key aspects of the human disease. Without a thorough understanding of the disease, the mode of action of the drug candidate and species-specific physiology and pathology the development of a disease model that mimics the human condition is difficult. Continuous cell lines derived from cancer patients have widely been used as in vitro models to study the biology of specific tumor types and response to treatment. There are several broad panels of cancer cell lines available, many of them with extensive genomic characterization and drug cytotoxicity profiles. These panels include diverse cancer cell panels, cancer-specific collections e.g. breast cancer and colorectal cancer cell line panels as well as engineered isogenic cancer cell lines. The most widely used panel of cell lines is the **NCI-60 cancer cell line panel**. In the 1980s the National Cancer Institute (NCI) assembled a collection of 60 cell lines from nine different human malignancies and obtained extensive data on genomics, transcriptomics, proteomics, metabolomics and drug response (Gillet et al. 2013). The collected data sets are publicly available and represent very useful tools to study the effect of a drug candidate on a specific cancer type or the correlation between treatment response and the profile of somatic mutations. Data obtained from the NCI-60 panel which contain only a limited number of lines from a specific cancer type can be extended by using homogeneous cancer cell line models that might be capable of predicting the clinical response to treatment in a particular cancer type with a higher degree of accuracy. The use of human cell line models allows for the tight control and monitoring of experiments and is of crucial importance to investigate the mode of action of a drug candidate. These models, however, are constrained by the fact that cultured cells do not represent an accurate model of the complex human disease. Data obtained using human cell lines have to be validated in animal models, samples from human patients or in clinical trials. Mice are the most widely used animals to study efficacy in a preclinical setting. Mice share a 99% of genes with humans, they are easy to handle, relatively cheap and genetic is well established manipulation in this species. **Xenograft models** refer to the transplantation of cells or tissues from one species to another. As this procedure would lead to a rejection of the transplant in animals with an intact immune system, these experiments are performed in immunodeficient mice such as nude, severe combined immunodeficiency (SCID), NOD/SCID or NOD/SCID/gamma mice which display different degrees of

4.4 Pre-clinical Drug Development

immunodeficiency. We distinguish cell line-derived xenograft (CDX) and patient-derived xenograft (PDX) (Fig. 4.14).

The CDX model is based on the injection of cancer cells and the monitoring of the growth of the established tumor and the survival of the animals. Cancer cells can be either injected subcutaneously or implanted into the organ of origin. Later procedure is called orthotopic grafting and is considered as more clinically relevant. The size of the tumor can be determined by caliper measurements along the longest two dimensions of the tumor. Alternatively, the ectopic expression of luciferase in the transplanted cells allows to monitor tumor growth by using bioluminescence imaging. In order to assess the efficacy of the drug candidate to reduce tumor growth in vivo, tumor-bearing mice are treated with the compound and curves for tumor growth and survival are established. CDXs have been considered as artificial for several reasons including their incapacity of modelling the tumor environment. Conversely, **PDX systems** use tissue from the primary tumor of a patient to graft it without any intermediate in vitro culture step directly into the immunodeficient mouse (Hidalgo et al. 2014). PDX systems more accurately mimic the cell heterogeneity and the microenvironment of a tumor in the human body. However, both CDX and PDX lack the capacity to determine the contribution of the immune system in the response to exper-

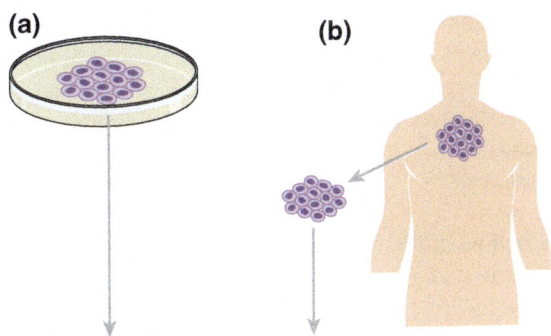
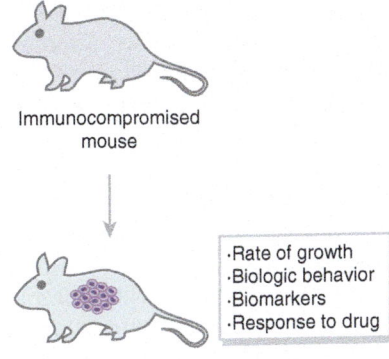

Fig. 4.14 Different mouse xenograft models. Cell line-derived (**a**) or patient-derived (**b**) cells can be transplanted into immunocompromised mice to study the effect of the drug candidate on tumor growth or survival

imental treatments. As there is solid evidence that the immune system significantly contributes to the clinical response to a broad range of cancer therapies, the use of immunodeficient mouse models raise serious concerns and is completely inadequate for the preclinical development of immunotherapeutics. The efficacy of drugs that act through activating the host immune system can be evaluated by using **immunocompetent mouse allograft models**. These models are based on the implantation of cells from a tumor of an inbred mouse strain into mice of the same inbred strain. Because the implanted cells and the host are genetically similar or identical, these models are called syngeneic. Importantly, the allografted cells can produce tumors in the presence of full murine immunity and comprehensive stroma. Major drawbacks of this approach are the limited number of available syngeneic cell lines and the limitation to murine biology which is not always predictive in the human context. **Genetically engineered mice (GEM)** used in preclinical drug development are the result of genetic manipulation introducing genetic alterations found in human cancer (Fig. 4.15; Richmond and Su 2008).

Tumors develop orthotopically through a process of tumor formation and progression in their native microenvironment and in the presence of full murine immunity. Despite the fact that these tumors are of murine origin and grow in a murine context, GEM models are thought to provide more clinically relevant data in preclinical efficacy testing of drug candidates. In order to generate mouse models that recapitulate features of human physiology, **humanized mouse models** have been developed (Brehm et al. 2010). These animals have been manipulated to carry human genes, cells or tissues. For example, mice bearing mutations in the IL2 receptor common gamma chain (IL2rgnull) are severely immunodeficient and can be used as recipients of human hematopoietic stem cells that develop into functional human immune systems.

Preclinical Pharmaceutics Evaluation—Can it be Delivered
The best drug is useless if it cannot reach the target tissue. Many drug candidates still have liabilities concerning solubility and bioavailability entering preclinical testing. Pharmaceutical formulation can help to improve drug pharmacokinetics such as water solubility and stability. Pharmaceutical formulation combines different chemical substances to a medication product that contains the active ingredient and excipients. Excipients have no medicinal activity but promote lubricity, flowability, disintegration and taste and facilitate absorption. In order to choose the right formulation approach, physicochemical characterization of the drug candidate in an industrial setup often begins in parallel with early preclinical evaluation. The preferred formulations for in vivo preclinical evaluation are solutions with co-solvents such as ethanol, propylene glycol, and polyethylene glycols where the drug candidate is presented already in a state suitable for absorption. If very high concentrations of the drug candidate have to be administered for example in in vivo toxicity evaluation suspensions can be used. The bioavailability in the gastrointestinal tract of poorly soluble drug candidates with ionizable groups can be enhanced several fold by salt formation. Formulation studies start in the preclinical development phase of the drug candidate but in general have not been concluded by the time it is evaluated in

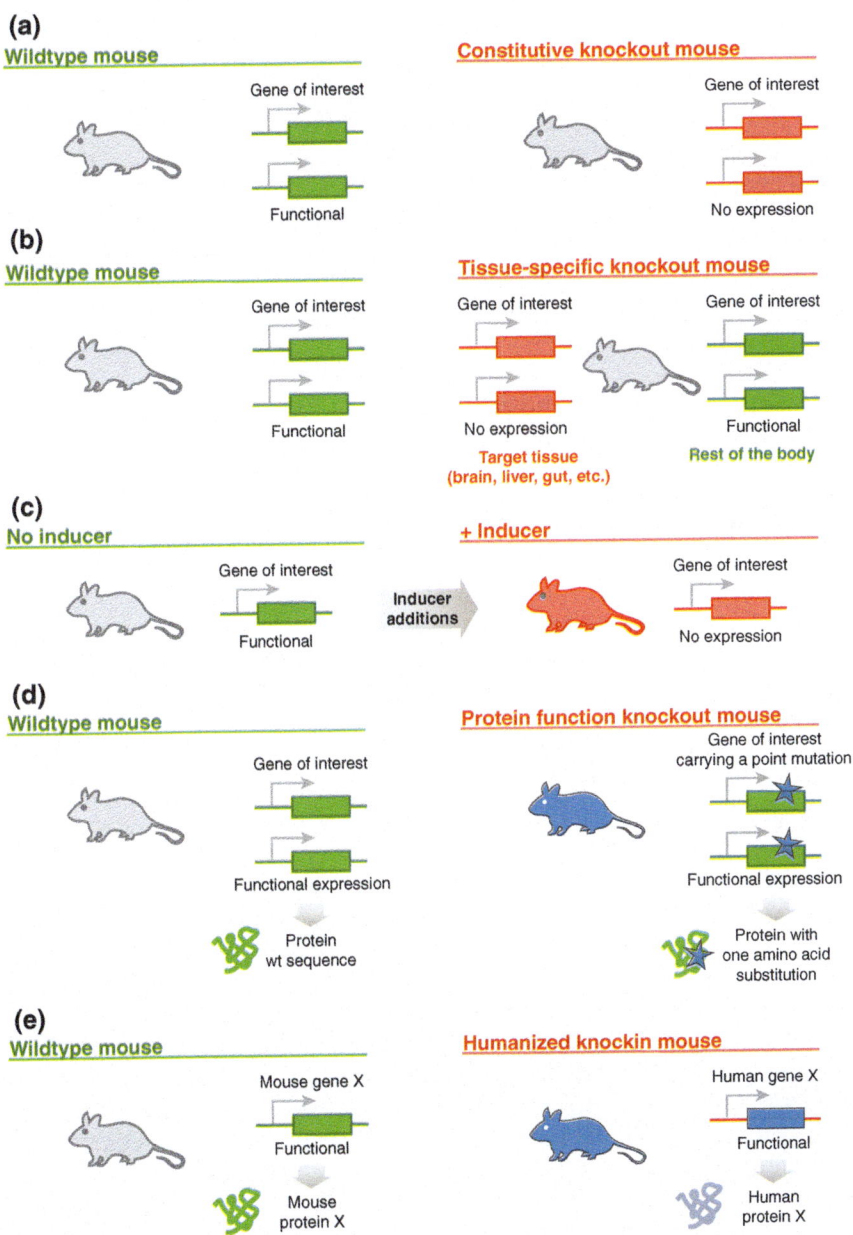

Fig. 4.15 Genetically engineered mouse models. Modern genetic engineering allows to generate sophisticated mouse models in which the molecular target of interest can be manipulated in a constitutive (**a**), tissue-specific (**b**) or inducible manner (**c**). The target of interest can be modified to contain the human mutation (**d**) or the mouse can be engineered to express the human gene (**e**)

clinical phase 1. Therefore, simple formulations of the drug candidate are initially used in early phases of human testing. At this stage long term stability evaluation of the formulation is not required. Once the drug candidate reaches phase 3 evaluation, pharmaceutical formulation should be defined and its long-term stability and the effect of factors such as temperature, humidity and light assessed. The field of drug delivery is a very active research field and many innovations are on the horizon. Targeting specific site in the human body and delivering drugs to specific cells improving efficacy while reducing systemic toxicity is a major challenge of drug delivery research. The emerging nanotechnology has an important impact in the field as nanoscale drug carriers have shown the potential to address some of these challenges.

Preclinical Evaluation of Synthetic Viability—Is its Manufacturing Viable
Library compounds, hit compounds and drug leads are usually made in a chemistry lab in small quantities. Techniques for synthesizing a chemical agent at a small scale do not translate always to larger production. During preclinical development, researchers must find a suitable strategy how to make large enough quantities of the drug for clinical trials. Clinical testing of a drug candidate might require tens of kilograms of the product. Therefore, the synthetic route must be suitable for scaling up the procedure. Limiting factors are structural complexity and size of the drug candidate. An important aspect is the reproducibility of the procedure and the potential for impurities and contaminants. Sometimes it might be useful to employ microbial processes to produce highly functional intermediates which then can be used to enhance the efficacy of the synthetic process. In the preclinical phase and even when clinical evaluation of a drug candidate has already started, the manufacturing process and analytical methods for quality control are still under development. Stability studies to define a suitable expiry date are still to be performed in parallel to the clinical trial. However, clinical evaluation of a drug candidate requires already a tightly controlled quality control system to ensure the quality and reproducibility of the resulting data. The batch size will be defined according to the size of clinical study.

Investigational New Drug (IND) Application
Before any drug candidate can be evaluated in humans, the researchers must file an Investigational New Drug (IND) application with a regulatory agency such the **Food and Drug Administration** (FDA) in the United States or the **European Medicines Agency** (EMA) in the European Union. An IND application provides the chemical structure of the drug candidate, the results of the preclinical studies on pharmacokinetics, efficacy and safety, the mode of action, list of side effects and manufacturing details. The sponsor has to demonstrate the capacity to adequately produce and supply consistent batches of the drug. The IND application also includes detailed information about the planned clinical trial outlining how, where and by whom the studies will be performed. The application contains information on the qualifications of clinical investigators to ensure that they are qualified to run the trial. The regulatory agency reviews the application to ensure that people participating in the clinical trials will not be exposed to unreasonable risks. Clinical trials are regulated

4.4 Pre-clinical Drug Development

by law and need to be approved by an **ethics committee** in Europe or an **institutional review board** (IRB), an which are independent committees of physicians, nurses, statisticians, community advocates and others at the institutions where the clinical trials will be performed has to approve the trial. The function of these committees is to make sure that a clinical trial is ethical and the right welfare of study participants is protected. A very important aspect of any clinical trial is the development of an appropriate **informed consent** document. Informed consent is a process for getting permission before conducting a healthcare intervention on a person. A signed informed consent will be required of all clinical trial participants. The informed consent informs about key aspect of the clinical trial including study purpose, risks and benefits, alternative treatments, confidentiality of the data and very importantly a statement that the participation is voluntary. The regulatory agency has thirty days to review the IND application. The rate of approval is about 88%.

Clinical Development

A drug candidate undergoes extensive evaluation in humans to prove that it is safe and therapeutically effective before the regulatory agency grants its approval for the drug to be used in clinical practice (Rubin and Gilliland 2012). Clinical trials are conducted on new drugs, but also on existing medicines which are repurposed for a new indication. This procedure involves a series of clinical phases each associated with specific questions (Fig. 4.16) objectives and requirements (Table 4.6).

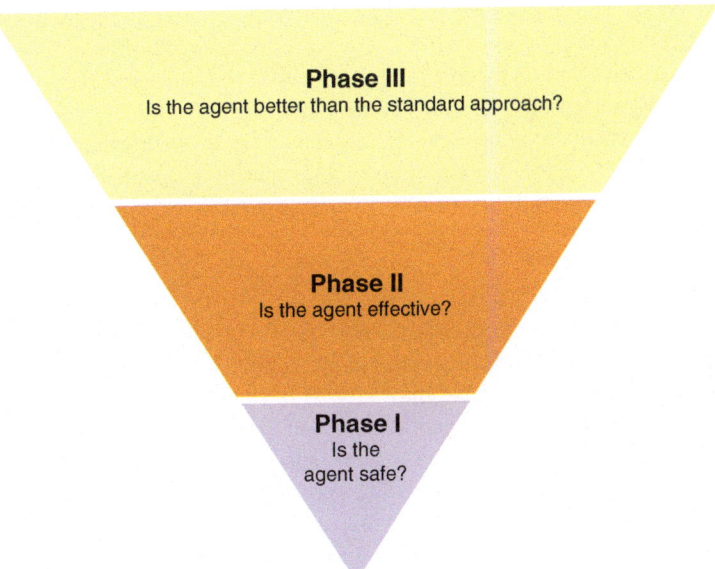

Fig. 4.16 Questions to be addressed in the three phases of clinical evaluation of a drug candidate before approval

Table 4.6 The different phases of clinical trials

Phase	Participants	Objectives
I	20–80	Evaluate Toxicity, ADME, Defining dose for phase II
II	100–400	Evaluate efficacy against tumors
III	1000–4000	Compare efficacy and safety with standard of care
IV	>5000	Post-marketing surveillance studies to identify rare adverse effects

New medicines usually undergo 3 phases of clinical trial. Phase 1 analyses the safety of the drug, phase 2 tests efficacy and finally phase 3 study safety and efficacy with a number of patients that allow statistically valid conclusions. After the approval, when the drug is already marketed post marketing surveillance is conducted. These studies are also called phase 4 trials and are aimed at analysing rare or long-term effects of the drug in a large population of patients.

Clinical trials serve as the basis for evidence-based medicine. A clinical trial is a complex procedure in which different actors have to work closely together including the clinical research team comprising physicians, nurses and statisticians, the sponsor, in most of the cases a pharmaceutical company, the regulatory agency and last but not least the patients. The sponsor or the regulatory agency can stop the trial at any time in case an unacceptable risk emerges or when the investigational drug performs extremely well and therefore it would be unethical to exclude patients that receive a placebo or another less efficient drug from the therapeutic benefits of the new drug candidate. The sponsor has to comprehensively report to the regulatory agency and the IRB on the progress of the clinical trial on a regular basis. Clinical trials are very expensive and time consuming lasting 6–7 years. Therefore, it is crucial to carefully design each clinical trial and to maximize the confidence in the validity of the results. There are many different aspects to be considered and depending on the characteristics of the disease, the research questions, therapeutic options, the endpoints and the availability of a control group many experimental options are available.

The Design of Clinical Trials

In order to maximize the success of a clinical trial it is essential to select the appropriate cohorts of patients. **Eligibility criteria** for a clinical trial can range from general to very specific requirements and may vary with trial phase. General eligibility criteria include age and sex of the participants and the type of cancer to be evaluated. The design of the trial often requires more specific criteria to be applied such as prior treatment, tumor characteristics e.g. specific mutation or disease stage, blood cell counts and organ function. If previous anti-cancer therapy is allowed the type of these treatments has to be clearly specified in the protocol. **Preselection** of trial participants according to molecular parameters is of particular relevance for targeted therapies that often rely on the presence of a specific molecular target. If the target of a targeted therapy is present only in a very small subpopulation of patients with a specific cancer type, evaluating this drug candidate without prior **biomarker-driven**

4.4 Pre-clinical Drug Development

stratification of possible trial participants according to the presence or absence of the target, will fail to prove clinical efficacy. An example for this scenario is the clinical evaluation of Crizotinib. As mentioned above only 5% of NSCLC patients carry the fusion gene EML4-ALK whose protein product can be targeted by the tyrosine kinase inhibitor Crizotinib. This drug was evaluated in preselected trial participants positive for EML4-ALK. The response rate to Crizotinib treatment in this clinical trial was almost 60%. However, it also became clear that without biomarker based selection of the appropriate participants, this trial would have failed.

Clinical trials can involve one or several groups of patients receiving a specific treatment, usually referred to as trial arms. The simplest design of a clinical trial is the **single-arm trial** in which all patients receive the same experimental therapy. Therefore, single-arm trials do not compare data of patients treated with the experimental drug with patients that did not receive the drug or were treated with placebo or an alternative therapy. The patients are observed and treatment response and clinical outcome are monitored. Single arm trials are used to obtain preliminary evidence for the efficacy of a drug candidate or additional data on safety or when it is impossible to recruit a sufficient number of participants with a specific cancer to establish a control arm. Single-arm trials, also called non-randomized trials are often used for phase 1 or phase 2 trials in oncology because they are faster, cheaper and can provide evidence of efficacy. Single-arm trials have to be carefully designed to show efficacy in the absence of direct control data. As cancer is not a self-limiting disease, an improvement of the condition without treatment is not expected. Hence, comparing a specific parameter at the beginning and over the treatment period might provide evidence of efficacy. Alternatively, data obtained from a single-arm trial can be compared to published data on treatment of similar patients who did not receive the treatment or received a different treatment. However, the conditions under which these trials have been conducted are rarely comparable. Alternatively, a crossover study can be conducted in which each patient receives both treatments, the new drug and the control, in a randomized order so that the outcome for each treatment can be compared within each participant. However, **randomized placebo controlled clinical trials** provide much more solid data on safety and efficacy and are usually required for phase 3 trials before a new drug can be approved. Randomized clinical trials randomly distribute patients in different groups which then are treated with different drugs/therapies to compare their clinical outcomes. In a randomized clinical trial, an experimental arm, which receives the drug candidate, is compared with control arms, which receive another therapy e.g. the standard of care for the specific condition, a placebo or no treatment. A randomized patient only receives one treatment/or treatment combination during the duration of the trial. The parameters to be observed and compared between the groups have to be defined before the trial starts and the sample size has to be calculated according to requirements to obtain statistically useful data. The statistical power of a trial increases with the number of subjects enrolled but it is also getting more expensive and for some diseases it is not easy to recruit a great number of patients. Many clinical trials are designed as **single-blind or double-blind trials** in which the patients or the patients and the researchers do not know which treatment they are receiving or giving, respectively. In contrast,

Table 4.7 List of endpoints frequently used to measure safety and efficacy of a drug candidate in clinical trials

	Endpoint	Definition
OS	Overall survival	Time from randomization until death from any cause
PFS	Progression free survival	Time from randomization until disease progression or death
TTP	Time to progression	Time from randomization until tumor progression
ORR	Overall response rate	Proportion of patients with a predefined tumor size reduction
DoR	Duration of response	Time from tumor response to disease progression

in **open label trials** both the patients and the researchers know which treatment is being administered. Although the blinded setting is less biased, sometimes blinding is impossible as it is obvious to the participants in which arm of the trial they have been placed in. Safety and efficacy of a drug candidate in clinical trials are measured by using specific outcomes, called **endpoints** (Table 4.7; Fleming and Powers 2012). These endpoints have to be predefined before the clinical trial is conducted and can include **clinical endpoints** or **surrogate endpoints**. Clinically meaningful endpoints directly measure how a patient feels, functions, or survives e.g. heart failure, while a surrogate endpoint is a laboratory value or a physical sign that is faster and easier to determine and used to predict a clinically meaningful endpoint e.g. blood pressure. The primary endpoint is the main parameter used to measure the clinical benefit of the new treatment. Although **overall survival** (OS) defined as the time from beginning of the clinical trial until death from any cause e.g. survival duration in month, is the most important clinical outcome and the preferred endpoint as it represents the most important benefit for cancer patients. In addition, OS is unambiguous and easy to assess. However, the use of OS as an endpoint is associated with significant limitations as it requires a large sample size and a long follow-up. As the power for the trial depends on the adequate number of events (reached endpoints), OS as an endpoint may be problematic when median OS is relatively long. Results might be calculated once 50% of subjects have reached the endpoint or after all participants reached the endpoint.

The estimation of OS can also be confounded by subsequent therapies after progressive disease on trial treatments which are allowed for ethical reasons. Another commonly used endpoint is **Progression Free Survival** (PFS), the length of time during and after the treatment that a patient lives with the disease but it does not get worse. Similarly, **Time to Progression** (TTP) measures the time to objective tumor progression but does not include patients who die before the disease progresses. **Overall Response Rate** (ORR) determines the proportion of patients with tumor size reduction of a predefined amount and for a minimum time period and is commonly used as an endpoint in phase 2 trials. However, response rates are often temporary and may not translate into a longer overall survival of the patient. Even when the drug candidate leads to significant response rates and a prolonged disease-

4.4 Pre-clinical Drug Development

free survival but does not extend OS and is associated with significant side effect that impact the quality of life.

Phase 1 Clinical Trials

A phase 1 clinical trial is the first step in testing a new investigational drug or new use of a marketed drug in humans. This is the first time that the drug molecule is tested in humans. Therefore, phase 1 trials are a critical moment in the development of a new drug. This is where the rubber meets the road. Oncology phase 1 trials are typically single-arm, open-label, sequential studies that involve 20–80 patients with advanced cancer that has not responded to standard cancer treatments but with a good performance status and adequate organ function. It might be challenging to find fairly healthy advanced cancer patients. Therefore, clinical phase 1 trials sometimes recruit patients with different types of cancer. In phase 1 clinical studies are usually conducted open label and emphasis is put on drug safety. Phase 1 trials are exploratory, non-therapeutic from which the participants usually cannot expect a therapeutic benefit. Phase 1 clinical trials rather determine pharmacokinetic parameters and toxicity and establish a dose and/or schedule of a candidate drug for testing its efficacy in phase 2 trials. Phase 1 clinical trials include first-in-human trials and dose escalation trials. In order to test the drug candidate in humans for the first time, trial participants are divided into small groups, known as cohorts. The first cohort, usually 3–6 patients, receives a low dose of the new drug which is very unlikely to cause serious toxicity. Starting dose is one tenth of the lethal dose (LD10) as tested in the most sensitive species in the preclinical phase. In the absence of any major adverse side effects, the dose is escalated until pre-determined safety levels are reached, or intolerable side effects start showing up. Clinical signs are monitored and blood or urine samples might be collected to measure drug levels or biomarkers in the patients. The pharmacokinetics of the drug candidate are evaluated to determine how the it is absorbed, distributed, metabolized and excreted. Dose escalation studies are designed to establish the highest dose with acceptable side effects, known as the **maximum tolerated dose** (MTD). Drug induced toxicity is analyzed relative to the dose and unexpected side effects are explored. The side effects that determine the MTD are known as **dose-limiting toxicities** (DLTs) and must be defined before the trial begins e.g. grade 4 neutropenia lasting longer than 5 days, grade 4 thrombocytopenia, grade 3 non-heme toxicity. MTD is usually defined as the highest dose level at which one or more patients out of 6 develop a DLT. Furthermore, researchers characterize the metabolism, drug-drug interaction, food effects, QT interval of the electrocardiogram and routes of excretion of the candidate drug. Phase 1 clinical trials last about 1 year. About 70% of drugs pass this phase. In recent years, phase 0 or micro-dose studies are often performed prior to clinical phase 1 trials. Phase 0 trials are optional, exploratory studies in which 10–15 patients are exposed to a single sub-therapeutic dose to obtain pharmacokinetic data.

Phase 2 Clinical Trials

In phase 2 clinical trials, the candidate drug is tested to see if it has any therapeutic effect and to determine the dose needed for this effect. Phase 2 trials are **thera-**

peutic exploratory studies on a limited scale using a fixed dose of the drug candidate to evaluate its therapeutic efficacy. They typically involve 100–300 individuals who have the specific cancer type to be targeted. Based on data obtained from the preclinical evaluation, knowledge about the molecular mode of action of the drug candidate and clinical phase 1 data eligibility criteria for the recruitment to the phase 2 trials are defined. The selection of participants is aimed at the recruitment of a homogenous population likely to respond to the experimental treatment. Eligibility criteria might include a specific molecular profile of the tumor or limit the number of prior treatments. Phase 2 trials may be conducted at several different medical centers or hospitals to enhance the capacity of recruiting suitable participants. This design is referred to as **multicenter trials**. The choice of the target cancer type is based on data obtained from pre-clinical experiments and phase 1 clinical trials that can provide information on the mode of action of the drug candidate and the most suitable indication. As the success of targeted anticancer treatments depends on the presence of a specific molecular target, the selection of suitable patients is key for testing these agents in phase 2 clinical trials. An increasingly important aspect in phase 2 trials for targeted agents is the development of mechanism-based biomarker to determine if the candidate drug affects the intended target. Phase 2 trials are often conducted as single-arm trials comparing the obtained data with historic data from published trials that established the efficacy of another treatment. Some phase 2 trials are designed to compare different dosing arms within the trial. However, the growing number of approved anti-cancer treatments facilitates their use as comparators in randomized phase 2 clinical trials. Clinical phase 2 trials are often subdivided into phase 2a and phase 2b. Phase 2a trials refer to pilot studies on a smaller scale to provide proof of concept for clinical efficacy of the drug candidate while phase 2b is usually aimed at finding the right dosing to achieve the best efficacy and minimizing side effects. However, the limited number of patients and safety concerns usually prevent phase 2 trials from establishing efficacy of the experimental treatment. However, phase 2 trials are essential to prepare a careful design of the subsequent phase 3 clinical trial including finding the optimal dose and scheduling, administration routes, and endpoints. Phase 2 clinical trials last about 2 years and may cost 20 million USD. Of all the clinical phases, phase 2 clinical trials have the lowest success rate. Only about 30% of drug candidates advanced from Phase 2 to Phase 3. Upon conclusion of phase 2 clinical trials the regulatory agency, the sponsor and the investigator review the preliminary data and assess the viability of progressing the drug candidate further to a phase 3 clinical trial.

Phase 3 Clinical Trials

Phase 3 clinical trials are **therapeutic confirmatory** studies on large number of patients to demonstrate that the candidate drug works. Phase 3 clinical trials typically involve 1000–3000 patients. Phase 3 trials are based on data from prior clinical studies demonstrating safety and potential efficacy of the drug candidate. In order to generate statistically significant data about safety and efficacy phase 3 clinical trials are often conducted as multi-center trials, in order to enhance recruitment of a large and diverse target population. In order to ensure statistical power of the study, sample size is carefully calculated considering the drop-outs from the trial. Similar to phase 2 trials, phase 3 clinical studies have well-defined eligibility criteria. In order control for confounding, stratification strategies are often applied to ensure a balanced representation of patient parameters among the treatment arms. Phase 3 clinical trials should characterize the effect of the candidate drug in different populations considering patient variations in genetics, life style and concomitant conditions such as liver impairment or pregnancy using different dosages as well as combined treatment with other drugs. Phase 3 trials also determine the incidence of common adverse reactions. The most commonly used endpoints to evaluate efficacy of the drug candidate is OS. PFS is often used as a surrogate endpoint to predict OS in a procedure for accelerated approval followed by evidence from OS for full approval. Serious adverse events, dose adjustments, lab parameter values and vital signs might be used as **safety endpoints**. If possible, this phase is designed as randomized (patients are randomly allocated to receive one or other of the alternative treatments) and double-blind (neither the participants nor the researchers know who is receiving a particular treatment) controlled studies. In a comparative efficacy or **superiority trial** which is the most common type of phase 3 clinical trials, the drug candidate is compared with existing treatments focused on safety and efficacy. Other types of phase 3 clinical trials such as **equivalency trials** or **non-inferiority trials** are designed to determine whether the treatment with the drug candidate is similar to a **comparator drug**. These types of studies are commonly used to evaluate biosimilars. Phase 3 clinical trials should confirm therapeutic efficacy in the target population and determine the safety profile. It also provides the basis for labeling instructions to ensure proper use of the drug. Phase 3 clinical trials last about 3 years and may cost as much 53 million USD. Due to the duration and large number of participants, phase 3 clinical trials are the most expensive clinical studies and pose significant challenges for adequate trial design. Often, two different successful phase 3 trials are required to obtain regulatory approval. Upon conclusion of the phase 3 evaluation, the results of all the studies performed including preclinical and clinical phases are put together in a large document called **New Drug Application** (NDA) and submitted to the corresponding regulatory agency. About 50% of drugs that are evaluated in phase 3 trials fail or are rejected by the regulatory agency. In order to allow patients access to the experimental treatment until the approved drug can be purchased, some phase 3 trials continue while the regulatory agency reviews the NDA. The sponsors commonly use this type of studies which are often referred to as phase 3b trials to obtain evidence that the drug candidate provides benefits for patients with additional types of tumors facilitating label expansion.

4.5 Approval Process for a New Drug

All new drugs have to be approved by regulatory authorities such as the Food and Drug Administration (FDA) in the United States or the European Medicines Agency (EMA) in the European Union governing the respective market before a particular product can be introduced into the market (Liberti et al. 2010). These agencies evaluate new drugs based on the evidence presented from the clinical studies. In the US, these data are provided by the sponsor in the so called "New Drug Application" (NDA). After NDA approval is obtained, the pharmaceutical company will market the drug. To be approved, a new drug has to be non-inferior or better than an approved drug. Non-inferior outcome ensures that a survival advantage associated with an approved drug will not be lost with a new agent. A NDA contains a comprehensive description of all the data gathered during the different phases of preclinical and clinical development of the drug. The application includes the methods and results of human and animal studies, patent information, manufacturing procedures, possible abuse, compliance and conclusions of completed clinical trials by the Institutional Review Board, formulation details, shelf life and proposed labeling. The purpose of a NDA is to demonstrate that a drug is safe and effective for its intended use in the population studied. Once submitted, the regulatory agency has 60 days to perform a preliminary review. The review team can refuse to file the NDA if it is found incomplete. If the team decides that the NDA is acceptable for full review a standard or an accelerated review procedure will be applied which implies a final decision within ten or six months, respectively. Sometimes, the regulatory authority asks to address remaining issues or to conduct additional studies. The decision whether to approve or not to approve the drug will be communicated in a complete response letter. A critical aspect of this decision process is the balance between benefits and risks associated with the new drug. Furthermore, the regulatory agency decides what information is needed to add to the package insert to guide the appropriate use of the new drug. Biologics which are considered as more complex than chemical drugs require a **Biologic License Application (BLA)** a procedure that is different from a NDA. The FDA has mechanisms for accelerating the review and approval process for promising drugs, including fast track designation, accelerated approval, and priority review. **Fast track** is a specific procedure to facilitate the development of drugs to treat serious conditions for which no or only limited treatment options exist. A drug candidate that receives fast track designation is eligible for several privileges such as a rolling review process which implies evaluation of parts of the still incomplete NDA or BLA, accelerated approval and priority review. A **priority review** designation can be given for drugs that treat diseases without therapy or drugs that provide major benefits over existing treatments or as an incentive to develop drugs for rare and neglected diseases. **Accelerated approval** designation allows for the use of surrogate endpoints to predict clinical benefit. As surrogate endpoints require less time approval can be achieved faster. However, drug approved under the accelerated procedure still are required to demonstrate clinical benefits in phase 4 clinical trials. FDA can withdraw approval for drug that fail to show these benefits.

Phase 4 Clinical Trials

After a drug has been approved and introduced into the market often phase 4 clinical trials are conducted. Phase 4 trials are **postmarketing surveillance** studies or confirmatory trials sometimes requested by the regulatory authority or performed by the sponsor for competitive or safety reasons (Fontanarosa et al. 2004). Phase 4 trials is

4.5 Approval Process for a New Drug

the moment when the approved drug is tested for the first time in the real world. Phase 3 clinical trial commonly enroll no more than 3000 participants and therefore are unable to reflect the diversity of populations. The statistical power of phase 3 trials is limited. Adverse reactions that occur in fewer than 1 in 3000 patients are unlikely to be detected in Phase 3 clinical trials. Additional, limitation of phase 3 trials are their relatively short duration and strict inclusion/exclusion criteria. But in the real world no patient can be excluded. Even patients with co-morbidities such as liver of kidney dysfunction or pregnant woman have to be treated. Less-common adverse drug reactions can be identified monitoring a much larger patient population for a longer time period after marketing of the drug. Accordingly, about 20% of drugs acquire black box warnings after marketing. A black box warning is the strictest warning issued by FDA to alert patients and physicians to potential life-threatening, or very serious, side effects. Approximately 4% of already approved drugs are ultimately withdrawn for safety reasons. Phase 4 trials are commonly designed as open-label multi-center trials conducted in varied populations. The minimum time period mandatory for Phase 4 trials is two years.

Thought Questions

1. What is meant by a lead compound in medicinal chemistry?

    (A) A drug containing the element lead
    (B) A leading drug in a particular area of medicine
    (C) A compound that acts as a starting point for drug development
    (D) A drug which is normally the first to be described for a particular disease/aliment

2. Which of the following statements is one of the Lipinski's rules (Rule of Five)?

    (A) An orally active drug has a molecular weight equal to 500
    (B) An orally active drug has no more than five hydrogen bond acceptor groups
    (C) An orally active drug has no more than 10 hydrogen bond donor groups
    (D) An orally active drug has a calculated logP value less than +5

3. Which of the following objectives in drug development is not related to pharmacodynamics?

    (A) The reduction of side effects
    (B) The optimization of activity
    (C) The reduction of toxicity
    (D) The maximization of oral bioavailability

4. Pharmacokinetics is defined as
    (A) The study of biological and therapeutic effects of drugs
    (B) The study of absorption, distribution, metabolism and excretion of drugs
    (C) The study of mechanisms of drug action
    (D) The study of methods of new drug development

5. Which of the following types of clinical trials determines whether a targeted therapy works against cancer?

   (A) Phase I
   (B) Phase II
   (C) Phase III
   (D) Phase II and Phase III
   (E) Phase I, Phase II, and Phase III.

References

Benson JD et al (2006a) Validating cancer drug targets. Nature 441:451–456. https://doi.org/10.1038/nature04873
Brehm MA, Shultz LD, Greiner DL (2010) Humanized mouse models to study human diseases. Curr Opin Endocrinol Diabetes Obes 17:120–125. https://doi.org/10.1097/MED.0b013e328337282f
Cancer Genome Atlas Research N et al (2013) The cancer genome atlas pan-cancer analysis project. Nat Genet 45:1113–1120. https://doi.org/10.1038/ng.2764
Drews J (2000) Drug discovery: a historical perspective. Science 287:1960–1964
Eglen RM, Reisine T, Roby P, Rouleau N, Illy C, Bosse R, Bielefeld M (2008) The use of AlphaScreen technology in HTS: current status. Curr Chem Genom 1:2–10. https://doi.org/10.2174/1875397300801010002
Faller B (2008) Artificial membrane assays to assess permeability. Curr Drug Metab 9:886–892
Ferreira BI, Hill R, Link W (2015) Special review: caught in the crosshairs: targeted drugs and personalized medicine. Cancer J 21:441–447. https://doi.org/10.1097/PPO.0000000000000161
Fleming TR, Powers JH (2012) Biomarkers and surrogate endpoints in clinical trials. Stat Med 31:2973–2984. https://doi.org/10.1002/sim.5403
Fontanarosa PB, Rennie D, DeAngelis CD (2004) Postmarketing surveillance–lack of vigilance, lack of trust. JAMA 292:2647–2650. https://doi.org/10.1001/jama.292.21.2647
Garraway LA, Sellers WR (2006) Lineage dependency and lineage-survival oncogenes in human cancer. Nat Rev Cancer 6:593–602. https://doi.org/10.1038/nrc1947
Gillet JP, Varma S, Gottesman MM (2013) The clinical relevance of cancer cell lines. J Natl Cancer Inst 105:452–458. https://doi.org/10.1093/jnci/djt007
Hajduk PJ, Greer J (2007) A decade of fragment-based drug design: strategic advances and lessons learned. Nat Rev Drug Discov 6:211–219. https://doi.org/10.1038/nrd2220
Hargrave-Thomas E, Yu B, Reynisson J (2012) Serendipity in anticancer drug discovery World. J Clin Oncol 3:1–6. https://doi.org/10.5306/wjco.v3.i1.1
Hewitt RE (2011) Biobanking: the foundation of personalized medicine. Curr Opin Oncol 23:112–119. https://doi.org/10.1097/cco.0b013e32834161b8
Hidalgo M et al (2014) Patient-derived xenograft models: an emerging platform for translational cancer research. Cancer Discov 4:998–1013. https://doi.org/10.1158/2159-8290.CD-14-0001
Hodgson J (2001) ADMET–turning chemicals into drugs. Nat Biotechnol 19:722–726. https://doi.org/10.1038/90761
Hughes JP, Rees S, Kalindjian SB, Philpott KL (2011a) Principles of early drug discovery. Br J Pharmacol 162:1239–1249. https://doi.org/10.1111/j.1476-5381.2010.01127.x
Kaelin WG Jr (2005) The concept of synthetic lethality in the context of anticancer therapy. Nat Rev Cancer 5:689–698. https://doi.org/10.1038/nrc1691
Kim JH, Scialli AR (2011) Thalidomide: the tragedy of birth defects and the effective treatment of disease. Toxicol Sci 122:1–6. https://doi.org/10.1093/toxsci/kfr088

References

Knowles J, Gromo G (2003) A guide to drug discovery: target selection in drug discovery. Nat Rev Drug Discov 2:63–69. https://doi.org/10.1038/nrd986

Kostiainen R, Kotiaho T, Kuuranne T, Auriola S (2003) Liquid chromatography/atmospheric pressure ionization-mass spectrometry in drug metabolism studies. J Mass Spectrom 38:357–372. https://doi.org/10.1002/jms.481

Lanthier M, Behrman R, Nardinelli C (2008) Economic issues with follow-on protein products. Nat Rev Drug Discov 7:733–737. https://doi.org/10.1038/nrd2636

Lee WM (2003) Drug-induced hepatotoxicity. N Engl J Med 349:474–485. https://doi.org/10.1056/NEJMra021844

Li AP (2004) Accurate prediction of human drug toxicity: a major challenge in drug development. Chem Biol Interact 150:3–7. https://doi.org/10.1016/j.cbi.2004.09.008

Liberti L, Breckenridge A, Eichler HG, Peterson R, McAuslane N, Walker S (2010) Expediting patients' access to medicines by improving the predictability of drug development and the regulatory approval process. Clin Pharmacol Ther 87:27–31. https://doi.org/10.1038/clpt.2009.179

Lindsay MA (2003) Target discovery. Nat Rev Drug Discov 2:831–838. https://doi.org/10.1038/nrd1202

Lipinski CA, Lombardo F, Dominy BW, Feeney PJ (2001) Experimental and computational approaches to estimate solubility and permeability in drug discovery and development settings. Adv Drug Deliv Rev 46:3–26

Mavromoustakos T, Durdagi S, Koukoulitsa C, Simcic M, Papadopoulos MG, Hodoscek M, Grdadolnik SG (2011) Strategies in the rational drug design. Curr Med Chem 18:2517–2530

Nebert DW, Russell DW (2002) Clinical importance of the cytochromes P450. Lancet 360:1155–1162. https://doi.org/10.1016/s0140-6736(02)11203-7

Pai VB, Nahata MC (2000) Cardiotoxicity of chemotherapeutic agents: incidence, treatment and prevention. Drug Saf 22:263–302. https://doi.org/10.2165/00002018-200022040-00002

Perkins R, Fang H, Tong W, Welsh WJ (2003) Quantitative structure-activity relationship methods: perspectives on drug discovery and toxicology. Environ Toxicol Chem 22:1666–1679

Pushpakom S et al (2018) Drug repurposing: progress, challenges and recommendations. Nat Rev Drug Discov. https://doi.org/10.1038/nrd.2018.168

Richmond A, Su Y (2008) Mouse xenograft models vs GEM models for human cancer therapeutics. Dis Model Mech 1:78–82. https://doi.org/10.1242/dmm.000976

Rubin EH, Gilliland DG (2012) Drug development and clinical trials-the path to an approved cancer drug. Nat Rev Clin Oncol 9:215–222. https://doi.org/10.1038/nrclinonc.2012.22

Sams-Dodd F (2005) Target-based drug discovery: is something wrong? Drug Discov Today 10:139–147. https://doi.org/10.1016/S1359-6446(04)03316-1

Santos R et al (2017) A comprehensive map of molecular drug targets. Nat Rev Drug Discov 16:19–34. https://doi.org/10.1038/nrd.2016.230

Sheiner LB, Steimer JL (2000) Pharmacokinetic/pharmacodynamic modeling in drug development. Annu Rev Pharmacol Toxicol 40:67–95. https://doi.org/10.1146/annurev.pharmtox.40.1.67

Smith C (2003) Drug target validation: hitting the target. Nature 422:341, 343, 345 passim. https://doi.org/10.1038/422341a

Smith DA, Di L, Kerns EH (2010) The effect of plasma protein binding on in vivo efficacy: misconceptions in drug discovery. Nat Rev Drug Discov 9:929–939. https://doi.org/10.1038/nrd3287

Sneader W (2000) The discovery of aspirin: a reappraisal. Bmj 321:1591–1594

Terstappen GC, Schlupen C, Raggiaschi R, Gaviraghi G (2007) Target deconvolution strategies in drug discovery. Nat Rev Drug Discov 6:891–903. https://doi.org/10.1038/nrd2410

Walsky RL, Obach RS (2004) Validated assays for human cytochrome P450 activities. Drug Metab Dispos 32:647–660. https://doi.org/10.1124/dmd.32.6.647

Waring MJ et al (2015) An analysis of the attrition of drug candidates from four major pharmaceutical companies. Nat Rev Drug Discov 14:475–486. https://doi.org/10.1038/nrd4609

Weinstein IB, Joe A (2008) Oncogene addiction. Cancer Res 68:3077–3080; discussion 3080. https://doi.org/10.1158/0008-5472.can-07-3293

Zanella F, Lorens JB, Link W (2010) High content screening: seeing is believing. Trends Biotechnol 28:237–245. https://doi.org/10.1016/j.tibtech.2010.02.005

Zhang JH, Chung TD, Oldenburg KR (1999) A simple statistical parameter for use in evaluation and validation of high throughput screening assays. J Biomol Screen 4:67–73. https://doi.org/10.1177/108705719900400206

Further Reading

Benson JD, Chen YNP, Cornell-Kennon SA, Dorsch M, Kim S, Leszczyniecka M, Sellers WR, Lengauer C (2006) Validating cancer drug targets. Nature 441:451–456

Eric H, Rubin EH, Gilliland DG (2012) Drug development and clinical trials—the path to an approved cancer drug. Nat Rev Clin Oncol 9:215–222

Gibbs JB (2000) Mechanism-based target identification and drug discovery in cancer research. Science 287:1969–1973

Hughes JP, Rees S, Kalindjian SB, Philpott KL (2011) Principles of early drug discovery. Br J Pharmacol 162:1239–1249

Kola I, Landis J (2004) Can the pharmaceutical industry reduce attrition rates? Nat Rev 3:711–715

Rho JP, Louie SG (2003) Handbook of pharmaceutical biotechnology. Pharmaceutical Products Press, New York

Sawyers C (2004) Targeted cancer therapy. Nature 432:294–297

van't Veer LJ, Bernards R (2008) Enabling personalized cancer medicine through analysis of gene-expression patterns. Nature 452:564–570

Chapter 5
Economic and Social Implications of Modern Drug Discovery

As it has been mentioned earlier, we live exciting times in cancer research and within the near future, we will have efficient treatment options for many types of human cancers. However, these new medicines come with a significant price tag. The price of cancer drugs increased enormously within the last 15 years and a treatment course with a targeted anticancer drug can easily cost $100,000 (Morgan et al. 2011). The prices for new cancer drugs will soon become unaffordable for our societies and will limit the access to these therapies to the wealthy (Mazzucato 2016). Therefore, it is worth to discuss the economic and social implications of modern drug development.

The discovery and development of new medicines is a lengthy process, associated with an extremely high risk of failure and in turn very expensive. It takes an average of 12–15 years to evaluate an experimental drug in preclinical and clinical tests and to get one medicine to market eventually. The expenses associated with research and development for one approved drug which includes costs for failed drugs total between $100 million an $1000 million. The significant discrepancy between these figures is due to estimates that are based on different parameters e.g. focussing on new molecular entities (see Box 4 in Chap. 4) which require the most expensive type of research or on less innovative drugs, or subtracting the expenses that are tax deductible. In any case, it is very expensive. But why does it matter how expensive it is to develop a drug. It is important because pharmaceutical companies justify high prices with their investments in research and development of the drug and its benefit to patients. However, if survival or quality of life were taken as the criteria to measure benfit, there is certainly no correlation between price and benefits. Many targeted therapies only prolong life by a few month or weeks and are still very expensive. It has been claimed that restraining prices for prescription drugs would jeopradize investment in the research on future medicines. However, there is yet another important aspect to take into account. Taxpayer-funded scientists laid the groundwork for the discovery and development of most new cancer medicines. The discovery of drugs in the past required less molecular understanding than the identification of modern targeted drugs which rely on the previous identification and validation of critical molecular targets (Ferreira et al. 2015). Innovative drug discovery needs well-validated targets. Therefore, target identification/validation is the single most important step in the innovation value chain. As innovation is intrinsically prone to failure, this early phase of the drug discovery process is the lengthiest and riskiest one. Remarkably, as target identification/validation is mainly based on taxpayers-funded research, the state is the main risk taker in this phase. The development of Imatinib, the paradigmatic example for an innovative targeted drug was based on several key enabling scientific discoveries in publicly-funded research labs. In 1960 Peter Nowell and David Hungerford at University of Pennsylvania and Fox Chase Cancer Center in Philadelphia, respectively discovered the Philadelphia chromosome, an acquired chromosome translocation associated with chronic myelogenous leukemia (CML) (Nowell and Hungerford 1960). In 1973, Janet D. Rowley at the University of Chicago identified the mechanism by which the Philadelphia chromosome was formed (Rowley 1973) and between 1981 and 1987 Nora Heisterkamp and John Groffen at the National Cancer Institute together with collogues from the Erasmus University in Rotterdam, Netherlands found out that the inter-chromosomal exchange gives rise to the BCR-ABL fusion whose constitutive tyrosine kinase activity was shown to drive CML by work carried out at UCLA. The identification and validation of BCR-ABL as a therapeutic target enabled Brian Druker, at Oregon Health and Science University and Nicholas Lydon, at the pharmaceutical company Novartis to develop a specific inhibitor against it. Imatinib is not an exception (de Klein et al. 1982). It is rather a typical example of modern day targeted drug discovery. Almost all innovative new medicines that have been approved in recent years trace their research to publicly-funded research labs. Interestingly, the contribution of publicly-funded research correlates with the degree of therapeutic improvement of a new drug over existing drugs. The input from publicly-funded research labs is particularly important for the NCEs which represent the most innovative molecules. It is important to note that the majority of new drugs approved for sale are not NMEs whose number has been steadily declining over the past 15 years. In order to avoid risks and to compete with products that already proved to be clinically useful, the pharmaceutical industry outsourced innovative drug research and invested in the development of me too drugs (see Box 4 in Chap. 4). As a consequence, the risk/reward ratio for pharmaceutical innovation is unbalanced between the public and the private sector (Link 2018). Indeed, taxpayers pay twice, through publicly subsidized research and then to get access to overprized medicines. Therefore, changing the distribution of benefits from innovation would promote scientific discoveries for future anti-cancer therapies and ensure the access to the fruits of the outstanding cancer research conducted during the last decades.

References

de Klein A et al (1982) A cellular oncogene is translocated to the Philadelphia chromosome in chronic myelocytic leukaemia. Nature 300:765–767

Ferreira BI, Hill R, Link W (2015) Special review: caught in the crosshairs: targeted drugs and personalized medicine. Cancer J 21:441–447. https://doi.org/10.1097/PPO.0000000000000161

References

Link W (2018a) Knowledge-based drug discovery intensifies private appropriation of publicly financed research. Lancet Oncol 19:1017–1018. https://doi.org/10.1016/S1470-2045(18)30437-6

Mazzucato M (2016) High cost of new drugs. Bmj 354:i4136. https://doi.org/10.1136/bmj.i4136

Morgan S, Grootendorst P, Lexchin J, Cunningham C, Greyson D (2011) The cost of drug development: a systematic review. Health Policy 100:4–17. https://doi.org/10.1016/j.healthpol.2010.12.002

Nowell PC, Hungerford DA (1960) A minute chromosome in human chronic granulocytic leukemia. Science 132

Rowley JD (1973) Letter: a new consistent chromosomal abnormality in chronic myelogenous leukaemia identified by quinacrine fluorescence and Giemsa staining. Nature 243:290–293

Further Reading

Baker D (2017) Drugs are cheap: why do we let governments make them expensive?. Uppsala University, The Svedberg Seminar

Galkina CE, Beierlein JM, Khanuja NS, McNamee LM, Ledley FD (2018) Contribution of NIH funding to new drug approvals 2010–2016. Proc Natl Acad Sci 115:2329–2334

Light DW, Warburton R (2011) Demythologizing the high costs of pharmaceutical research. BioSocieties 6:34–50

Link W (2018b) Knowledge-based drug discovery intensifies private appropriation of publicly financed research. Lancet Oncol 19:1017–1018

Mazzucato M (2013) The entrepreneurial state: debunking public vs private sector myths. Anthem Press, London

Index

A

Absorption, Distribution, Metabolism, Excretion (ADME), 107, 109, 110, 116
Aclarubicin, 26, 27
Acute lymphoblastic leukemia, 4, 15, 52, 53, 57, 58, 64, 67
Adjuvant treatment, 7, 8, 55
Adverse effect, 119
Afatinib, 37, 38
Alectinib, 37, 39
Alemtuzumab, 52, 53
Alkylating agents, 15–18, 31, 32, 80
Amrubicin, 27, 28
Anaplastic large cell lymphoma, 57, 58
Anaplastic Lymphoma Kinase (ALK), 35, 37, 39, 80, 127
Anemia, 30
Angiogenesis, 32, 40, 55, 56, 92, 99
Anthracyclines, 15, 26–28, 31, 117
Antibody
   based drugs, 51
   dependent cellular cytotoxicity, 51, 53
   drug conjugates, 56, 91
   enzyme conjugate, 58
   humanized, 49, 53–55, 58
   monoclonal, 10, 11, 34, 38, 40, 49, 52, 54, 72, 77
   radioisotope conjugate, 58
Antigen
   binding region, 49
   binding site, 50, 53
   oncofetal, 61
   presenting cells, 65, 67, 68
   tumor-associated, 61
   tumor-specific, 61
Anti-metabolites, 15

Aromatase inhibitors, 4, 35
Assay
   Alpha Screen, 101
   biochemical, 100, 101, 105
   Caco-2 permeability, 109, 111
   cell-based, 93, 94, 100, 105, 111
   cell-free, 100
   equilibrium dialysis, 111
   hERG, 117
   In vitro, 89, 108, 109, 113, 114, 116
   metabolic stability, 108, 109, 113
   multiplexed, 101
   PAMPA, 109
   ultrafiltration, 112
Atezolizumab, 69
ATP-binding cassette, 78, 79
Avelumab, 69

B

Basal cell carcinoma, 46, 47, 49
Base excision repair, 44, 45
B cells, 11, 52, 53, 58, 60, 61, 67
Belinostat, 43, 44
Bevacizumab, 53, 55, 56
Bioavailability, 98, 109, 122, 133
Biologic license application, 132
Bleomycin, 27–29
Blinatumomab, 52, 53
Blood
   brain barrier, 17, 18, 31, 108, 111
   cell, 3, 14, 29, 30, 126
   clotting, 30
   flow, 111, 115
   forming tissues, 3
   plasma, 111
   samples, 97

stream, 14, 108, 111, 115
supply, 12
testis barrier, 111
vessel, 39, 40, 115
Bortezomib, 44, 48, 49
Brachytherapy, 10
BRCA1/2, 12, 44, 45
Brentuximab vedotin, 57, 58

C
Cancer-Associated Fibroblasts (CAFs), 83
Cancer stem cells, 2, 12, 32, 46
Carboplatin, 16–18, 82
Carcinoma, 2, 3, 24, 40, 53, 56, 69, 83, 117
Cardiotoxicity, 27–29, 31, 116, 117
Carfilzomib, 49
Carmustine, 17
Cell therapy
  adoptive, 64
  CAR-T, 66
  Dendritic, 60, 64–66, 68, 71
Ceritinib, 37, 39
Cetuximab, 53, 55
Chemotherapy, 4, 7, 8, 13–15, 18, 24, 29–32, 38, 55, 56, 58, 71, 78
Chidamide Epidaza, 43, 44
Chimeric Antigen Receptor (CAR), 66, 67
Chronic Myelogenous Leukaemia (CML), 37, 40, 91
Cisplatin, 4, 16–18, 20, 30, 31, 82
Clinical trial
  double-blind, 127, 131
  equivalency, 131
  multicenter, 130
  non-inferiority, 131
  open label, 127, 129
  Phase I, 127, 129, 130
  Phase II, 113, 125, 127–130
  Phase III, 38, 104, 127, 130–133
  Phase IV, 126, 132, 133
  placebo controlled, 127
  randomized, 127, 131
  superiority, 131
Colorectal cancers, 55
Companion diagnostics, 32, 41, 42
Compound
  chemical, 15, 18, 21, 22, 24, 38, 88, 99, 101, 103, 110
  collection, 24, 103
  library, 38, 103, 124
  screening, 42, 100, 103, 104
Copanlisib, 37, 43

CRISPR, 93, 94
Crizotinib, 37, 39, 42, 80, 127
CTLA-4, 51, 67–70
Cutaneous T cell lymphoma, 44, 52
Cyclopamine, 46, 47
Cyclophosphamide, 17, 30, 79, 80
Cytarabine, 19, 20, 80
Cytochromes P450, 114
Cytokine, 62–64, 66, 71
Cytotoxic antibiotics, 15, 26, 27

D
Dabrafenib, 37, 41, 42
Dacarbazine, 17, 18, 71
Daratumumab, 52
Dasatinib, 37, 41, 59
Daunorubicin, 27
Dendritic
  cell therapy, 64, 66
Dexamethasone, 44, 49
Diffuse large B-cell lymphoma, 57, 58, 64
Dinutuximab, 52
Distribution, 29, 98, 102, 107, 109, 111, 133, 138
DNA
  double-strand breaks, 11, 25
  repair, 12, 16, 21, 44, 45, 91
DNA-PK, 12
Docetaxel, 23, 24, 55
Dose
  lethal, 129
  limiting toxicity, 29, 30, 48
  maximum tolerated, 119, 129
  response, 104
Doxorubicin, 27, 28, 31, 71, 80
Drug
  attrition, 92, 98, 103–105, 109, 116, 117
  candidate, 89, 92, 98, 105–107, 112, 113, 116–122, 124–132
  design, 20, 39, 99
  likeness, 106, 107
  potency, 104, 105, 108
  repurposing, 103, 104
  selectivity, 105
Drug development
  clinical, 48, 89, 99
  pre-clinical, 89, 107, 118, 119, 122
  process, 99, 103, 106, 117, 118
Drug discovery
  fragment-based, 41, 103
Druggability, 33, 92
Drug resistance

acquired, 32, 39, 41, 42, 59, 77
  cell autonomous, 77
  intrinsic, 77
  non-cell-autonomous, 77
Durvalumab, 69

E

Elotuzumab, 52, 53
Epidermal Growth Factor Receptor (EGFR), 35, 37, 38, 51, 53–55, 80, 83
Epithelial-to-mesenchymal transition, 83
Erlotinib, 37–39, 78, 80, 82
Estrogen receptor-α, 35
Etoposide, 24, 26, 27
European Medicines Agency (EMA), 124, 131
Everolimus, 37, 43
Excretion, 98, 107–109, 112, 115, 129, 133
Extracellular matrix, 58, 83, 95

F

Fibroblast Growth Factor Receptor (FGFR), 38
Fludarabine, 19, 21
Fluorescence In Situ Hybridization (FISH), 55
Fluorouracil, 19, 20, 71, 79
Follicular lymphoma, 37, 43, 52, 57, 58, 64
Food and Drug Administration (FDA), 4, 26, 28, 31, 33, 39, 42, 55, 99, 124, 131–133

G

Gastrointestinal Stromal Tumor (GIST), 37, 40, 41
Gefitinib, 37–39, 80, 82
Gemcitabine, 19, 20, 79
GLI, 46
Good laboratory practice (GLP), 118
Granulocyte-macrophage colony stimulating factor, 66

H

Head and neck cancer, 53, 71
Hematopoietic malignancies, 2, 3
Hepatotoxicity, 29, 31, 116, 117
Herpes simplex virus 1, 71
High-performance liquid chromatography (HPLC), 110
Histone acetyltransferases, 43, 44

Histone deacetylases, 35, 43, 44
Hit
  compound, 90, 100, 104–106, 108, 124
  expansion, 104, 108
  to lead, 104
Homologous recombination repair, 45
Human Epidermal Growth Factor Receptor 2 (HER2), 4, 37, 38, 45, 51, 54, 55, 57–59
Hyperthermia, 8, 9, 13

I

Ibritumomab tiuxetan, 11, 57, 58
Idelalisib, 37, 43
Imatinib, 4, 17, 34, 37, 40–42, 59, 78, 80–82, 91, 138
Immune
  checkpoint therapy, 62, 67, 68
  surveillance, 62
  system, 1, 2, 4, 30, 49, 57, 59–63, 65, 67, 68, 70, 120–122
Immunoediting, 62
Immunohistochemistry, 35, 55, 97
Immunotherapy
  active, 62, 63
  passive, 62–64
Immunotoxins, 34, 51, 56, 57, 77
Informed consent, 125
Inotuzumab ozogamicin, 57, 58
Institutional review board, 125, 132
Intellectual property, 99, 105, 106
Interferon-α, 64
Interleukin-2, 57, 64
Investigational New Drug (IND), 118, 124, 125
Ipilimumab, 68, 69
Irinotecan, 26
Ixazomib, 49

K

Kinase inhibitors, 34, 35, 37, 38, 40, 51, 80, 82, 105

L

Lapatinib, 37, 38
Lead
  identification, 89
  optimization, 47, 89, 98, 107–109, 113, 116, 117
Lipinski's Rule of Five, 106

Liquid chromatography/Mass spectrometry (LC/MS), 110
Liver extracts
   microsome, 114
   S9, 114
Lomustine, 17
Lymphoma, 11, 15, 20, 57, 58, 67, 69

M

Major histocompatibility complex, 61, 65, 66
Mantle cell lymphoma, 49
MAPK, 41, 42, 81
MDR1, 78, 84
Mechlorethamine, 17
Medulloblastoma, 46, 47
Melanoma, 4, 18, 32, 37, 41, 42, 51, 61, 64, 68, 69, 71, 81–83
Merkel-cell carcinoma, 69
Metabolism
   Phase-I, 112
   Phase-II, 112
Methotrexate, 19, 20, 78
Methylguanine
   Methyltransferase (MGMT), 82, 84
Mitoxantrone, 26
Mouse models
   genetically engineered, 120, 123
   humanized, 122
   immunocompetent allograft, 122
   xenograft, 120, 121
MTOR, 37, 41, 43, 81
Multiple myeloma, 24, 44, 47–49, 52, 53
Myelosuppression, 27–30

N

Neoadjuvant treatment, 8
Neoantigen, 61
Neuroectodermal malignancies, 2, 3
New Chemical Entity (NCE), 99
New Drug Application (NDA), 131, 132
New Molecular Entity (NME), 99
Nilotinib, 37, 38, 41, 78
Niraparib, 45
Nivolumab, 69
Non-hodgkin's lymphoma, 52, 58
Non–small-cell lung cancer, 37, 38
Nucleotide Excision Repair (NER), 44, 82

O

Obinutuzumab, 52, 58
Ofatumumab, 52, 58
Olaparib, 42, 45
Olaratumab, 53, 56
Oncogene
   addiction, 91
Oncolytic virus, 8
Overall response rate, 128
Oxaliplatin, 17, 18

P

Paclitaxel, 23, 24, 71
Palliative treatment, 7, 13
Panitumumab, 53, 55
Panobinostat, 43, 44
Pathogen-associated molecular pattern, 65
Pattern Recognition Receptors, 65
PD-1, 51, 67–70
PDGFR, 51, 54
Pembrolizumab, 69
Pemetrexed, 19, 20, 80
Peripheral T-cell lymphoma, 44
Pertuzumab, 53, 55
Pharmaceutical
   formulation, 122, 124
   industry, 92, 109, 116, 138
   innovation, 99, 138
Pharmacodynamics, 107, 118, 133
Pharmacokinetics, 41, 107, 115, 118, 122, 124, 129, 133
Pharmacophore, 47, 108
PI3K, 12, 35, 41–43, 81, 82
PIP3, 42, 81
Plasma protein binding, 111, 112
Poly ADP Ribose Polymerases (PARP), 35, 44, 45, 91
Precision oncology, 32
Progression free survival, 128
PTCH1, 46
PTEN, 12, 82, 84

R

Radiation therapy, 8–13, 31, 45
Ramucirumab, 53, 55, 56, 91, 92
Rapamycin, 43
Reactive oxygen species, 11, 31
Receptor tyrosine kinase, 40
Renal cell carcinoma, 37, 40, 43, 64, 69
Rituximab, 52, 58, 91
Rituximab and hyaluronidase human, 57, 58, 92
Romidepsin, 43, 44
Rucaparib, 45

Index

S

Sarcoma, 3, 24, 27, 53, 56, 64
Screening
    compound, 42, 100, 104
    high Content, 101, 102
    high Throughput, 104
Severe combined immunodeficiency, 120
Side effect, 51, 104, 128
6-Mercaptpurine, 19
Sonic hedgehog (Shh), 46, 47
Small molecule inhibitors, 34, 40, 41, 43, 96
Smoothened Receptor (SMO), 35, 46, 47
Solubility, 26, 47, 105, 108, 109, 111, 112, 122
Sonidegib, 47
Sorafenib, 37, 38, 40, 92
Spindle poisons, 15, 21, 23, 59
Stem Cell growth Factor Receptor (SCFR), 41
Structure-activity-relationship, 108
Sunitinib, 37, 38, 40, 78, 92
Suppressor of Fused (SUFU), 46
Synthetic lethality, 45, 91

T

Target
    deconvolution, 90
    discovery, 92
    Identification, 50, 88, 89, 92, 99, 137, 138
    validation, 88–90, 92–97, 137, 138
Targeted
    gene knockin, 96, 97
    gene knockout, 96
    therapy, 32, 35, 49, 50, 53, 55, 72, 126, 134
Taxanes, 22–24, 30, 31, 78
T-cell
    Lymphoma, 44, 52, 57
    regulatory, 64
Temozolomide, 17, 18, 82, 84

Temsirolimus, 37, 43
Teniposide, 26, 27
Teratomas, 3
Thalidomide, 119
Therapeutic target, 33, 50, 51, 72, 77, 80, 92, 97, 138
Time to progression, 128
Topoisomerase inhibitors, 15, 24–27, 31, 32, 78
Topotecan, 26, 71
Trametinib, 37, 42, 51
Trastuzumab
    emtansine, 58, 59
Tumor
    antigen, 60–62, 64–67, 71
    associated macrophage, 83
    atypical, 2, 3
    heterogeneity, 83, 121
    microenvironment, 83, 92, 94, 95, 121, 122
    model, 94, 96
T-vec, 71

V

VEGF, 35, 39, 40, 51, 53, 55, 56, 92
Vemurafenib, 37, 41
Vinblastine, 23, 24
Vinca alkaloids, 23, 24, 30–32, 71, 72
Vincristine, 23, 24, 71
Vinorelbine, 23
Vismodegib, 47
Vorinostat, 43, 44

X

Xenograft
    cell line-derived, 121
    models, 120, 121
    patient-derived, 121

The manufacturer's authorised representative in the EU is Springer Nature Customer Service Centre GmbH, Europaplatz 3, 69115 Heidelberg, Germany. If you have any concerns regarding our products, please contact ProductSafety@springernature.com

Printed and bound by CPI Group (UK) Ltd, Croydon, CR0 4YY

23/03/2026

02076379-0002